Foundations of
Social Evolution

MONOGRAPHS IN
BEHAVIOR AND ECOLOGY

Edited by John R. Krebs and
Tim Clutton-Brock

Foundations of Social Evolution

STEVEN A. FRANK

Princeton University Press
Princeton, New Jersey

Copyright © 1998 by Princeton University Press
Published by Princeton University Press,
41 William Street,
Princeton, New Jersey 08540
In the United Kingdom: Princeton University Press,
Chichester, West Sussex

Library of Congress Cataloging-in-Publication Data

Frank, Steven A., 1957–
Foundations of Social Evolution / Steven A. Frank
p. cm. — (Monographs in behavior and ecology)
Includes bibliographic references (p.) and index.
ISBN 0-691-05933-0 (cloth: alk. paper)
ISBN 0-691-05934-9 (pbk.: alk. paper)
1. Natural selection. 2. Behavior evolution. 3. Kin
selection (evolution). 4. Social evolution—Economic
models. I. Title. II. Series.
QH375.F735   1998
576.8'2'011—dc21
97-52086 CIP

Typeset by the author with TeX
Composed in Lucida Bright

Princeton University Press books are
printed on acid-free paper and meet the guidelines
for permanence and durability of the Committee
on Production Guidelines for Book Longevity
of the Council on Library Resources

Printed in the United States of America

10 9 8 7 6 5 4 3 2 1

10 9 8 7 6 5 4 3 2 1
(Pbk.)

Contents

Preface

Social evolution occurs when there is a tension between conflict and co-operation. The earliest replicating molecules inevitably competed with their neighbors for essential resources. They also shared a common interest in using local resources efficiently; otherwise, more prudent cartels would eventually drive overly competitive groups out of business. The conflicts and shared reproductive interests among cells within a complex organism, or among members of a honey bee colony, also qualify as social phenomena.

This book is about the economic concepts of value used to study social evolution. It is both a "how to" guide for making mathematical models and a summary with new insight about the fundamentals of natural selection and social interaction.

I have cast the subject in a manner that is comfortable for an evolutionary biologist but retains sufficient generality to appeal to many kinds of readers. These include economists, engineers who use evolutionary algorithms, and those who study artificial life to gain insight about evolution, cognition, or robotics.

A fellowship from the John Simon Guggenheim Foundation in 1995–1996 allowed me to catch up on other work. Andrew Pomiankowski and Yoh Iwasa invited me to join them at the Institute for Advanced Study in Berlin in 1996-1997, which provided an ideal opportunity for writing. The National Science Foundation supported my research during this period. My wife, Robin Bush, listened patiently and advised wisely.

Foundations of
Social Evolution

1 Introduction

> The elder Geoffroy and Goethe propounded, at about the same
> time, their law of compensation or balancement of growth; or,
> as Goethe expressed it, "In order to spend on one side, nature
> is forced to economise on the other side."
> —Charles Darwin, *On the Origin of Species*

The theory of natural selection has always had a close affinity with economic principles. Darwin's masterwork is about scarcity of resource, struggle for existence, efficiency of form, and measure of value. If offspring tend to be like their parents, then natural selection produces a degree of economic efficiency measured by reproductive success. The reason is simple: the relatively inefficient have failed to reproduce and have disappeared.

This book is about the proper measure of value in economic analyses of social behavior. Some count of offspring is clearly what matters. But whose reproductive success should be measured? Three exchange rates define the value influenced by natural selection.

Fisher (1958a) formulated reproductive value by direct analogy with the time value of money. The value of money next year must be discounted by the prevailing interest rate when compared to money today. Likewise, the value of next year's offspring must be discounted by the population growth rate when compared with the value of an offspring today. This exchange rate makes sense because the ultimate measure of value is not number of offspring, but contribution to the future of the population. In general, individuals must be weighted by their expected future contribution, their reproductive value.

The second factor is marginal value. This provides the proper scaling to compare costs and benefits of different consequences on the same scale, as in all economic analyses.

These first two exchange measures are standard aspects of economics and biology. The third scaling factor is defined by the coefficient of relatedness from kin selection theory. This exchange appears, at first glance, to be a special property of evolutionary analysis.

The theory of kin selection defines how an individual values the reproduction of a relative compared with its own reproduction (Hamilton 1964a). The following is a typical analysis. Sisters share by genealogical descent one-half of their genes. This relatedness coefficient of one-half means that natural selection is indifferent between a female who uses resources to produce one offspring of her own or gives those resources to her sister to produce two offspring. The *one-half* is an exchange rate for evolutionary value, because the same number of copies of a gene is made whether by one direct offspring, or by two indirect offspring each devalued by one-half.

Genealogy provides an appealing notion of kinship and value. However, Hamilton (1970) showed that kin selection properly values social partners according to statistical measures of genetic similarity that do not necessarily depend on genealogical kinship. This must be so because future consequences are determined only by present similarity, not by the past complexities of genealogy. The current theory of kin selection uses coefficients based on Hamilton's statistical measure of similarity.

Once one accepts statistical similarity as the proper measure of value, other puzzles arise, which have not been widely discussed. For example, interactions between different species are governed by the same form of statistical association as are interactions within species (Frank 1994a). But it does not make sense to speak of kinship or genetic similarity for interactions between species. Thus the simple notion of a genetic exchange rate in kin selection appears to be part of a wider phenomenon of correlated interaction.

I describe the current theory of kin selection in detail. I then show that kin selection has a close affinity to the ideas of correlated equilibrium in game theory and economics (Aumann 1974, 1987; Skyrms 1996). I connect these ideas to various notions of statistical information and prediction. This shows the logical unity of social evolution, statistical analysis of cause, aspects of Bayesian rationality, and economic measures of value.

I present the economic concepts of value by working through the methods needed to analyze particular problems. Thus the book also serves as a step-by-step guide for developing models of social evolution.

Chapter 2 is a self-contained summary of the main concepts and methods of analysis. This chapter also develops a statistical formalism of natural selection that detaches the theory from the particulars of genetics and biology. In spite of this abstraction, I preserve the language and style of typical biological models.

Chapter 3 reviews previous theories of kin selection. This chapter begins with Hamilton's (1970) derivation of inclusive fitness, which is a particular type of causal analysis for interactions among relatives. Queller's (1992a) model follows as an alternative to inclusive fitness, in which social interaction is analyzed as a problem in the evolution of correlated characters (Lande and Arnold 1983).

Chapter 4 develops new methods for studying social evolution. The first method extends Queller's analysis of correlated characters in social interaction. The second method transforms the analysis of correlated characters into an enhanced version of Hamilton's inclusive fitness theory. The measures of value are then used to develop maximization techniques. These techniques provide practical tools for solving problems.

Chapter 5 works through several cases of social interaction with correlated phenotypes. Many examples are in familiar game theory form. This provides background for the new interpretations of relatedness that follow.

Chapter 6 suggests that relatedness is, in fact, a statistical measure of information. Several examples are developed to illustrate this concept. This chapter also emphasizes the notion of conditional behavior, in which an individual adjusts its strategy in response to additional information. An example of kin recognition provides a natural connection between conditional behavior and the interpretation of relatedness as information.

Chapter 7 works through several examples of kin selection. Particular emphasis is placed on the distribution of resources and individuals. This shows how social behavior must be analyzed in its full ecological and demographic context. The models also illustrate how to use the techniques developed earlier to solve real problems.

Chapter 8 analyzes social interaction among different classes of individuals. The classes may be defined by age, size, or other attributes that change the marginal costs and benefits of sociality. A powerful technique is presented for combining class structure, reproductive value, and kin selection. The technique is illustrated by models of parasite virulence, social behavior in different kinds of habitats, and juvenile mortality in social groups. This chapter completes the presentation of fundamental principles and methods of analysis.

Chapters 9 through 11 summarize sex allocation theory. The problem of sex allocation is how a parent divides its resources between sons and daughters. The consequences depend on what neighbors do, creating a social aspect to payoffs for different strategies. Interactions among relatives change the shape of the payoffs. The analysis illustrates the methods and concepts of the previous chapters.

Chapter 12 reviews what has been accomplished and what remains to be done.

2 Natural Selection

"Natural selection is not evolution." This first sentence of Fisher's (1930) book, *The Genetical Theory of Natural Selection*, describes the limits of my analysis. I am concerned with the ways in which natural selection shapes patterns of biology. This is apart from many historical details of how evolution has actually proceeded. Suppose, for example, that humans were suddenly to become extinct. Perhaps another lineage of ape would follow the path of advancing language and intellect. Details of hair morphology, color, and development would likely differ from those of any extant people.

Maybe another humanlike lineage would never arise. The theory of natural selection is rather weak in predicting the special combination of ecological and genetic circumstances required to create a particular animal or plant. Rather, the theory is local. A question we might be able to answer is: for two otherwise similar populations that differ in a few parameters, what direction of change in social traits does the theory predict? This question emphasizes direction of change in a comparison. It is much more difficult to explain the degree of fit between organism and environment in a particular case.

So, in spite of *social evolution* in my title, this book is really about social natural selection. Even within this narrower scope, I have a limited goal. I am concerned with the logical deductions that follow from natural selection. I emphasize concepts and methods that aid rational thought, rather than an accounting of particular theories in light of the available data.

I begin with a summary of some useful tools for the analysis of natural selection. This summary provides an informal sketch of the principles. Later chapters fill in some of the formal detail and history for social topics.

2.1 Aggregate Quantities

> The regularity of [natural selection] is in fact guaranteed by
> the same circumstance which makes a statistical assemblage
> of particles, such as a bubble of gas obey, without appreciable
> deviation, the laws of gases.
> —R. A. Fisher, *The Genetical Theory of Natural Selection*

One of the first great challenges to the theory of natural selection came
with the rediscovery of Mendel's laws of heredity in 1900. Mendel showed that discrete characters, such as wrinkled or smooth peas, may each
be associated with a correspondingly discrete piece of hereditary ma-
terial. Each individual gets one hereditary particle for wrinkled, W, or
smooth, S, from each of its two parents. (Details of the following history
can be found in Provine (1971) and Bennett (1983).)

An offspring obtaining W from each parent, written as genotype WW,
expresses the wrinkled phenotype. An SS offspring is smooth. Another
of Mendel's interesting observations is that the mixed offspring, SW, is
smooth and phenotypically indistinguishable from the SS genotype. The
tendency of the mixed genotype to express the same phenotype as one
of the pure types is called dominance—one allele (hereditary particle) is
dominant over the other.

Mendel also studied the pattern of association between different characters. In one example, he analyzed simultaneously alternative colors,
white or yellow, and alternative textures, wrinkled or smooth. He found
that the color and texture qualities were inherited independently. Later
work showed that independent inheritance is common, but partial association (linkage) also occurs in many cases.

In summary, Mendel showed that characters are discrete with large
gaps between types, offspring with mixed hereditary particles express
one of the pure phenotypes rather than an intermediate character, and
different characters tend to be inherited independently.

The new Mendelians of the early 1900s interpreted these results as
a great challenge to the primacy of natural selection as an evolutionary
force. If there are large gaps between inherited characters, then differences between species must arise spontaneously by a new mutation of
large effect. This contradicts Darwin's emphasis on gradual change over

long periods of time—the slow and continuous reshaping of pattern by the inexorable process of selection acting on small variations.

The biometricians fought bitterly with the new Mendelians. They had been measuring the statistical properties of populations since Galton's work in the 1880s. Their characters, such as weight, varied continuously and were readily described by moments of distributions, such as means and standard deviations. Heredity was naturally described as the statistical association between parent and offspring—the correlation among relatives. Darwin's slow and continuous evolution by natural selection was readily understood by these statistical properties. Selecting heavier parents for breeding caused the mean weight of offspring to increase because parent and offspring are correlated. The increase was easily predicted by the excess among the selected parents and the parent-offspring correlation.

The biometricians did not have a theory that joined the facts of Mendelian heredity with the detailed observations of continuity and correlation. The Mendelians argued that, with the first clear information about heredity in hand, the case for particulate, discrete inheritance was settled and the biometricians' program was flawed. To resolve these opposing views, Yule suggested that many traits lacked the complete dominance found by Mendel. If many separate factors with incomplete dominance were combined, then Mendel's particulate inheritance might sum up to express continuous trait values.

Yule was correct but ignored. The turning point began with a public talk in 1911 by a young Cambridge undergraduate, R. A. Fisher. Fisher independently developed the idea that the dynamics of natural selection could be described by aggregate statistics of the hereditary particles. He explicitly discussed the analogy with statistical mechanics: the behavior of gases is often best described by the statistical properties of the population of molecules rather than the detailed dynamics of each particle. Fisher's (1918) classic paper settled the issue, although it took several years for the debate to subside.

Covariance of a Character and Fitness

Statistical descriptions of selection are useful for predicting short-term changes in populations. Such measures also provide a powerful method for reasoning about complex problems of selection independently of the underlying hereditary system.

The change in the average value of a trait is

$$\overline{w}\Delta\overline{z} = \text{Cov}\,(w,z) = \beta_{wz}V_z, \qquad (2.1)$$

where w is fitness and z is a quantitative character. I have assumed that z is inherited perfectly between parent and offspring. This assumption is relaxed below.

Eq. (2.1) shows that the change in the average value of a character, $\Delta\overline{z}$, depends on the covariance between the character and fitness or, equivalently, the regression coefficient of fitness on the character, β_{wz}, multiplied by the variance of the character, V_z. This equation was discovered independently by Robertson (1966), Li (1967), and Price (1970). The equation simply says that the more closely a character is associated with fitness, the more rapidly it will increase by selection.

Because fitness itself is a quantitative character, one can let the character z in Eq. (2.1) be equivalent to fitness, w. Then the regression, β_{ww}, is one, and the variance, V_w, is the variance in fitness. Thus the equation shows that the change in mean fitness, $\Delta\overline{w}$, is proportional to the variance in fitness, V_w. The fact that the change in mean fitness depends on the variance in fitness is usually called Fisher's fundamental theorem of natural selection, although that is not what Fisher (1958a, 1958b) really meant. Price (1972b) clarified Fisher's theorem in a fascinating paper that I will discuss later.

DYNAMIC SUFFICIENCY

The covariance equation can be thought of as a transformation from a system that specifies the dynamics of individual hereditary particles (alleles) to one that specifies the dynamics of the aggregate effects of the alleles. The transformed dynamics are expressed as the moments and cross-products of statistical properties of the population. This is useful because, in studying social evolution, the problem is to understand how selection changes the means and variances of social traits.

The great advantage of the covariance equation is that it depends only on relatively easily measured variables—the trait itself and the fitness of individuals with particular trait values. The cost for simplicity and few assumptions is that prediction of changes in the mean beyond the immediate response requires knowledge about the future value of the covariance. This is the problem of dynamic sufficiency. By making few assumptions about the dynamics of the causal particles, one can derive fewer consequences about system dynamics.

The conditions under which an evolutionary system is dynamically sufficient can be seen from the covariance equation, Eq. (2.1) (Frank 1995a). Initially, we require $\bar{z}$, $\bar{w}$, and $\overline{wz}$ to calculate $\Delta\bar{z}$ because $\text{Cov}(w, z) = \overline{wz} - \bar{w}\,\bar{z}$. We now have $\bar{z}$ after one time step, but to use the covariance equation again we also need $\text{Cov}(w, z)$ in the next time period. This requires equations for the dynamics of $\bar{w}$ and $\overline{wz}$. Changes in these quantities can be obtained by substituting either w or wz for z in Eq. (2.1); note that z can be used to represent any quantity, so we can substitute fitness, w, or the product of fitness and character value, wz, for z. The dynamics of $\overline{wz}$ are given by

$$\bar{w}\Delta\overline{wz} = \text{Cov}\,(w, wz) = \overline{w^2 z} - \bar{w}\,\overline{wz}.$$

Changes in the covariance over time depend on the dynamics of $\overline{wz}$, which in turn depends on $\overline{w^2 z}$, which depends on $\overline{w^3 z}$, and so on. Similarly, the dynamics of $\bar{w}$ depend on $\overline{w^2}$, which depends on $\overline{w^3}$, and so on. Dynamic sufficiency requires that higher moments can be expressed in terms of the lower moments (Barton and Turelli 1987).

Emphasis on the immediate (partial) direction of change caused by selection is a practical compromise to get a feeling for what is happening in complex systems. It requires careful use and study of limitations, as we will encounter in later sections.

Treating selection as a statistical process, based on aggregate quantities, is the first step toward a powerful method of analysis. The second step is partitioning evolutionary change into components, and assigning an explicit cause to each component.

2.2 Partitions and Causal Analysis

> [A] good notation has a subtlety and suggestiveness which at times make it seem almost like a live teacher.
>
> —Bertrand Russell

The evolutionary consequences of selection may be separated into different components. For example, inherited ability to withstand severe cold provides some individuals with a survival advantage during a winter storm. The adults that survive may differ from the population that existed previously. These changes are the direct effect of selection. The consequences for subsequent generations depend on the details of the inheritance system—the following generation is produced as the remaining adults breed and mix their alleles. The total change between generations can be partitioned into two components: the change among adults plus the change from breeding adults to offspring of the next generation.

A different partition emphasizes selection within and among groups. A selfish individual may outcompete its neighbors and increase its contribution to the next generation when compared with those neighbors. This is within-group selection. Selfish individuals may also reduce the efficiency and productivity of their group. The total contribution of a group with many selfish individuals will tend to be lower than the contribution of a group with few selfish individuals. Thus an individual's total success may be accounted by the combination of two levels: success relative to neighbors, and success of the neighborhood against other groups (Hamilton 1975; Wilson 1980).

Fisher's fundamental theorem of natural selection describes a third partition. Fisher separated changes in frequency caused directly by natural selection from other factors, which he called environmental effects. This theorem has been particularly confusing because Fisher ascribed indirect consequences of frequency change caused by natural selection to his second, environmental term. I will discuss this theorem below.

Partitions never change the total effect, and any total effect may be partitioned in various ways. Partitions are simply notational conventions and tools of reasoning. These tools may show logical connections and regularities among otherwise heterogeneous problems. Because alternative partitions are always possible, choice is partly a matter of taste. The possibility of alternatives leads to fruitless debate. Some authors inevitably claim their partition as somehow true; other partitions are labeled false when their goal or method is misunderstood or denigrated.

Statistical regression models, used for prediction or causal analysis, provide a complementary method of partition. Two steps have been particularly important in studies of natural selection. First, characters are described by their multiple regression on a set of predictor variables. The most common predictors in genetics are alleles and their interactions, but any predictor may be used. The second step is to describe fitness by multiple regression on characters. Once again, characters may be chosen arbitrarily.

Using these two steps in the Price Equation clarifies many historical aspects of the study of natural selection. In the following sections I show the relations among Fisher's fundamental theorem, Robertson's covariance theorem, the Lande and Arnold (1983) model for the causal analysis of natural selection, and Hamilton's rule for kin selection. This analysis not only unifies historical aspects, but also provides a powerful method for studying social evolution. (The following sections briefly summarize Frank (1997e), which provides additional details about partitions and causal analysis.)

The Price Equation

> Conceptual simplicity, recursiveness, and formal separation of levels of selection are attractive features of [Price's] equations.
> —W. D. Hamilton, "Innate Social Aptitudes of Man"

The Price Equation is an exact, complete description of evolutionary change under all conditions (Price 1970, 1972a). The equation adds considerable insight into many evolutionary problems by partitioning change into meaningful components.

Here is the derivation. Let there be a population (set) where each element is labeled by an index, i. The frequency of elements with index i is q_i, and each element with index i has some character, z_i. One can think of elements with a common index as forming a subpopulation that makes up a fraction, q_i, of the total population. No restrictions are placed on how elements may be grouped.

A second (descendant) population has frequencies q_i' and characters z_i'. The change in the average character value, $\bar{z}$, between the two populations is

$$\Delta\bar{z} = \sum q_i' z_i' - \sum q_i z_i. \tag{2.2}$$

Note that this equation applies to anything that evolves, since z may be defined in any way. For example, z_i may be the gene frequency of entities i, and thus $\bar{z}$ is the average gene frequency in the population, or z_i may be the square of a quantitative character, so that one can study the evolution of variances of traits. Applications are not limited to population genetics. For example, z_i may be the value of resources collected by bees foraging in the ith flower patch in a region (Frank 1997b) or the cash flow of a business competing for market share.

Both the power and the difficulty of the Price Equation come from the unusual way it associates entities from two populations, which are typically called the ancestral and descendant populations. The value of q_i' is not obtained from the frequency of elements with index i in the descendant population, but from the proportion of the descendant population that is derived from the elements with index i in the parent population. If we define the fitness of element i as w_i, the contribution to the descendant population from type i in the parent population, then $q_i' = q_i w_i / \overline{w}$, where $\overline{w}$ is the mean fitness of the parent population.

The assignment of character values z_i' also uses indices of the parent population. The value of z_i' is the average character value of the descendants of index i. Specifically, for an index i in the parent population, z_i' is obtained by weighting the character value of each entity in the descendant population by the fraction of the total fitness of i that it represents (see examples in later sections). The change in character value for descendants of i is defined as $\Delta z_i = z_i' - z_i$.

Eq. (2.2) is true with these definitions for q_i' and z_i'. We can proceed with the derivation by a few substitutions and rearrangements

$$\Delta \overline{z} = \sum q_i \, (w_i / \overline{w}) \, (z_i + \Delta z_i) - \sum q_i z_i$$
$$= \sum q_i \, (w_i / \overline{w} - 1) \, z_i + \sum q_i \, (w_i / \overline{w}) \, \Delta z_i,$$

which, using standard definitions from statistics for covariance (Cov) and expectation (E), yields the Price Equation

$$\overline{w} \Delta \overline{z} = \text{Cov}\,(w, z) + \text{E}\,(w \Delta z). \tag{2.3}$$

The two terms may be used to develop a variety of partitions because of the minimal restrictions used in the derivation (Hamilton 1975; Wade 1985; Frank 1995a). For example, the terms describe changes caused by selection and transmission, respectively. The covariance between fitness and character value gives the change in the character caused by differential reproductive success. The expectation term is a fitness-weighted measure of the change in character values between ancestor and descendant.

Recursive expansion of Eq. (2.3) provides another common partition (Hamilton 1975; Frank 1995a). For example, if the population is divided into groups, then w and z can denote the average fitness of a group and average character of a group, respectively. The covariance term then describes selection among groups. The expectation term subsumes selection within groups and other factors. This can be seen with self-expansion.

There is no satisfactory notation for hierarchical expansion. Each publication uses a different style to fit the particular problem. Here I use uppercase letters for individual values and lowercase for group means. Thus, for a particular group, $w = \overline{W}$ and $z = \overline{Z}$, so the expectation term is $E_G(\overline{W}\Delta\overline{Z})$. The subscript G emphasizes that the expectation is taken over groups. This form shows that the left side of the equation can be used to expand $\overline{W}\Delta\overline{Z}$, yielding

$$\overline{W}\Delta\overline{Z} = \mathrm{Cov}\,(W, Z) + \mathrm{E}\,(W\Delta Z), \qquad (2.4)$$

which expresses selection within the group in the covariance term and transmission in the expectation term. Since $w\Delta z = \overline{W}\Delta\overline{Z}$, Eq. (2.4) can be substituted into Eq. (2.3) to give the total change in the population

$$\overline{w}\Delta\overline{z} = \mathrm{Cov}\,(w, z) + \mathrm{E}_G\,[\mathrm{Cov}\,(W, Z) + \mathrm{E}\,(W\Delta Z)].$$

The equation could be used to expand the final term, $W\Delta Z$. Repeating the process provides an arbitrary number of hierarchical levels.

CAUSAL ANALYSIS

It is often convenient in measurement or in theoretical argument to consider explicitly the various factors that influence fitness. Multiple regression provides a useful set of tools to describe or estimate from data the direct effects of various predictors on fitness

$$w = \pi z + \beta'\mathbf{y} + \epsilon,$$

where π is the direct (partial regression) effect on fitness by the character under study, z, holding the other predictors $\mathbf{y} = (y_1, \ldots, y_n)^T$ constant, $\beta' = (\beta_1, \ldots, \beta_n)$ are the partial regression coefficients for the predictors, $\mathbf{y}$, and ϵ is the error in prediction.

Lande and Arnold (1983) analyzed natural selection and the change in character values within generations by study of

$$\overline{w}\Delta\overline{z} = \text{Cov}(w, z) = \pi\text{Cov}(z, z) + \sum_i \beta_i \text{Cov}(y_i, z).$$

This equation describes the direct effect of the character, z, on its own fitness, and the effect of correlated characters, $\mathbf{y}$, on the fitness of z. Expanded regression methods, such as path analysis, have been discussed widely (e.g., Crespi 1990). Heisler and Damuth (1987) and Goodnight et al. (1992) noted that one is free to use any predictors, $\mathbf{y}$, of interest. In particular, they emphasized that characteristics of groups can be used, allowing analysis of the direct effects of selection on group properties and the consequences for evolutionary change. I will return to this topic in a later section on kin selection.

Lande and Arnold (1983) extended their analysis to describe the response to selection, that is, the change in character values from one generation to the next. They used heritabilities to transform changes within a generation into changes between generations. However, heritabilities do not provide exact results when there is selection, even in theoretical models. I take an exact approach for character change between generations by using the Price Equation.

The difficulty for any method of describing character change between generations is that observed character values, z, will have many causes that are not easily understood. Further, some of those causes, such as random environmental effects, will not be transmissible to the next time period; thus Δz in the second term of Eq. (2.3) will be erratic and difficult to understand. It would be much better if, instead of working with z as the character under study, we could focus on those predictors of the character that can be clearly identified. It would also be useful if the transmissible properties of the predictive factors could be easily understood, so that some reasonable interpretation would be possible for Δz.

Let a set of potential predictors be $\mathbf{x} = (x_1, \ldots, x_n)^T$. Then any character z can be written as $z = \mathbf{b}'\mathbf{x} + \delta$, where the $\mathbf{b}'$ are partial regression coefficients for the slope of the character z on each predictor, x, and δ is the unexplained residual. The additive, or average, effect of each predictor, bx, is uncorrelated with the residual, δ.

In genetics the standard predictors are the hereditary particles (alleles). We write our standard regression equation for the character z of the ith individual as

$$z_i = \sum_j b_j x_{ij} + \delta_i = g_i + \delta_i, \qquad (2.5)$$

where $g_i = \sum_j b_j x_{ij}$, is the called the breeding value or additive genetic value. The breeding value is the best linear fit for the set of predictors, $\mathbf{x}_i$, in the ith individual. Each x_{ij} is the number of copies of a particular allele, j, in the ith individual. If we add the reasonable constraint that the total number of alleles per individual is constant, $\sum_j x_{ij} = K$, then the degree of freedom "released" by this constraint can be used among the b's to specify the mean of z. Thus, we can take $\bar{z} = \bar{g}$, and $\bar{\delta} = 0$.

The breeding value, g, is an important quantity in applied genetics (Falconer 1989). The best predictor of the trait in an offspring is usually $(1/2)(g_m + g_f)$, where g_m and g_f are genetic values of mother and father. Heritability is usually defined as V_g/V_z, where V_g is the variance in breeding values, g, and V_z is the variance in character values, z. There is, of course, nothing special about genetics in the use of best linear predictors in the Price Equation. The trait z could be corporate profits, with predictors, $\mathbf{x}$, of cash flow, years of experience by management, and so on. I will often use the term *allele* for *predictor*, but it should be understood that any predictor can be used.

A slightly altered version of Eq. (2.3) will turn out to be quite useful in the following sections. Any trait can be written as $z = g + \delta$, where g, the sum of the average effects, is uncorrelated with the residuals, δ. Average trait value is $\bar{z} = \bar{g}$, because the average of the residuals is always zero by the theory of least squares. In the next time, period $z' = g' + \delta'$, and $\bar{z}' = \bar{g}'$. Thus the change in average trait value is $\bar{z}' - \bar{z} = \Delta\bar{z} = \Delta\bar{g}$. To study the change in average trait value we need to analyze only $\Delta\bar{g}$, so we can use $z \equiv g$ in the Price Equation, yielding

$$\bar{w}\Delta\bar{z} = \bar{w}\Delta\bar{g} = \mathrm{Cov}\,(w, g) + \mathrm{E}\,(w\Delta g) \qquad (2.6)$$

$$= \beta_{wg}V_g + \mathrm{E}\,(w\Delta g), \qquad (2.7)$$

where, by definition of linear regression, $\mathrm{Cov}(w, g)$ can be partitioned into the product of the total regression coefficient, β_{wg}, and the variance, V_g, in trait value that can be ascribed to our set of predictors. In genetics, g is the (additive) genetic value, and V_g is the genetic variance. Yet another form of the Price Equation, obtained by simple rearrangement of terms in Eq. (2.7), will also turn out to be useful (Frank 1997e)

$$\Delta \overline{g} = \text{Cov}(w, g') / \overline{w} + \text{E}(\Delta g)$$
$$= \beta_{wg'} V_{g'} / \overline{w} + D_g. \tag{2.8}$$

Here g' is the breeding value transmitted by parents when measured among offspring. The first term accounts completely for differential fitness, and D_g is the change in breeding value between ancestor-descendant pairs.

Robertson (1966), in a different context, derived the $\text{Cov}(w, g)$ as the change in a character caused by natural selection. This covariance result is called Robertson's secondary theorem of natural selection, and is the form used by Lande and Arnold (1983) to describe evolutionary change between generations.

Robertson did not provide a summary of the remainder of total change not explained by the covariance term. Crow and Nagylaki (1976), expanding an approach developed by Kimura (1958), specified a variety of remainder terms that must be added to the covariance. They provided the remainders in the context of specific types of Mendelian genetic interactions, such as dominance, epistasis, and so on. The Price Equation has the advantages of being simple, exact, and universal, and we can see from Eq. (2.6) that, for total change, it is the term $\text{E}(w\Delta g)$ that must be added to the covariance term.

PREDICTORS AND ADDITIVITY

Confusion sometimes arises about the flexibility of predictors and of the Price Equation. The method itself adds or subtracts nothing from logical relations; the method is simply notation that clarifies relations. For example, in Eq. (2.5), I partitioned a character into the average, or additive, effect of individual predictors (alleles). One could just as easily study the multiplicative effect of pairs of alleles, including dominance and epistasis, by

$$z_i = \sum_j b_j x_{ij} + \sum_j \sum_k \alpha_{jk} x_{ij} x_{ik} + \delta_i = g_i + m_i + \delta_i,$$

where α_{jk} is the partial regression for multiplicative effects, and m_i is the total multiplicative effect of alleles. Then the analogous, exact expression for Eq. (2.6) is

$$\overline{w}\Delta\overline{z} = \overline{w}\Delta(\overline{g} + \overline{m}) = \text{Cov}(w, g + m) + \text{E}[w(\Delta g + \Delta m)].$$

Examples of the Price Equation applied to dominance and epistasis are in Frank and Slatkin (1990a). That paper showed how to calculate character change during transmission by direct calculation of $E[w(\Delta g + \Delta m)]$.

With respect to the general problem of additivity of effects, it is useful to recall the nature of least squares analysis in regression. That analysis makes additive the contribution of each factor, for example, $g + m$. But a factor, such as m, may be created by any functional combination of the individual predictors.

What is additivity? Unfortunately the term is used in different ways. Consider two contrasting definitions.

First, one can fit a partial regression (average effect) for each predictor in any particular population. The effects of the predictors can then be added to obtain a prediction for character value. Interactions among predictors (dominance and epistasis) can also be included in the model, and these partial regression terms are also added to get a prediction. The word *additivity* is sometimes used to describe the relative amount of variance explained by the direct effects of the predictors versus interactions among predictors.

Second, one can compare regression models between two different populations, for example, parent and offspring generations. If the partial regression coefficients for each predictor remain constant between the two populations, then the effects are sometimes called *additive*. This may occur because the context has changed little between the two populations, or because the predictors have constant effects over very different contexts.

Constancy of the average effects implies $E(w\Delta g) = 0$ in many genetical problems. This sometimes leads people to say that the equality requires or assumes additivity, but I find little meaning in that statement. Small changes in $E(w\Delta g)$ simply mean that the partial regression coefficients for various predictors have remained stable, either because the context has changed little or because the coefficients remain stable over varying contexts. Constancy may occur whether the relative amount of variance explained by the direct effects of the individual predictors is low or high.

Constancy of average effects plays an important role in a variety of well-known selection models. I use the Price Equation in the following sections to unify those models within a single framework.

FISHER'S FUNDAMENTAL THEOREM

R. A. Fisher stated his famous fundamental theorem of natural selection in 1930: "The rate of increase in fitness of any organism at any time is equal to its genetic variance in fitness at that time." He claimed that this law held "the supreme position among the biological sciences" and compared it with the second law of thermodynamics. Yet for 42 years no one could understand what the theorem was about, although it was frequently misquoted and misused to support a variety of spurious arguments (Frank and Slatkin 1992; Edwards 1994). Approximations and special cases were proved, but those sharply contradicted Fisher's claim of the general and essential role of his discovery. Price (1972b) was the first to explain the theorem and its peculiar logic. Price's work, known only to a few specialists, was clarified by Ewens (1989).

Price's (1970) own great contribution, the Price Equation, has a tantalizingly similar structure to the fundamental theorem, yet Price himself did not relate the two theories in any way. In this section I provide a proof of the fundamental theorem, following directly from the Price Equation (Frank 1997e). The proof combines the Price Equation with the models of causal analysis outlined in the previous sections.

Fisher did state that the rate of increase in the average fitness of a population is equal to the genetic variance in fitness. In spite of that statement, Fisher was not concerned with the total evolutionary change in fitness. Rather, he was interested in how natural selection directly changes the adaptation of individuals when studied in the context of total evolutionary change. By his definitions, natural selection inevitably increases individual fitness, but environmental changes act simultaneously in a way that usually reduces fitness by approximately the same amount. This must be so because, as Fisher noted, if average reproductive rate (fitness) were continually increasing or decreasing, then populations would either overrun the earth or quickly disappear.

By Fisher's view, the "partial" change in average fitness caused by natural selection is an increase proportional to the variance in fitness. The full evolutionary change in average fitness is the sum of the partial change in fitness caused by selection and a second term that is the partial change in fitness caused by changes in the environment. Environmental change includes every aspect of change in the genetic system, in interactions among individuals, and in the physical environment. Thus the natural selection term is extracted from the full evolutionary dynamics. The term focuses attention on selection as a single force in complex systems.

The Price Equation applies to general selective systems without any assumptions about the specifics of heredity. The equation has a similar, although not identical, partitioning between selective and environmental effects on evolutionary change. If, for example, we take fitness as the character under study, $z \equiv w$, then

$$\overline{w} \Delta \overline{w} = \mathrm{Cov}\,(w, w) + \mathrm{E}\,(w \Delta w)$$
$$= \mathrm{Var}\,(w) + \mathrm{E}\,(w \Delta w), \tag{2.9}$$

where the first term is the variance in fitness and the second is the component of evolutionary change caused by changes in the environment.

This is all a bit abstract. I show later how the partition between selective and environmental effects can be useful. The general point is that Eq. (2.9) provides an equilibrium condition

$$\mathrm{Var}\,(w) + \mathrm{E}\,(w \Delta w) = 0.$$

Selective improvements in fitness, $\mathrm{Var}(w)$, must be exactly balanced by what Fisher called "deterioration of the environment," here represented by $\mathrm{E}(w \Delta w)$.

Eq. (2.9) is similar to the fundamental theorem, but $\mathrm{Var}(w)$ is the total variance in fitness rather than Fisher's "genetic" variance. We can, however, prove the fundamental theorem directly from the Price Equation form given in Eq. (2.7). The trait of interest is fitness itself, $z \equiv w$, and, as for other traits, we write $w = g + \delta$. Thus $\beta_{wg} = 1$, and V_g is the genetic variance in fitness. Fisher was concerned with the part of the total change when the average effect of each predictor is held constant (Price 1972b; Ewens 1989). Since g is simply a sum of the average effects, holding the average effect of each predictor constant is equivalent to holding the breeding values, g, constant, thus $\mathrm{E}(w \Delta g) = 0$ (Frank 1997e). The remaining partial change is the genetic variance in fitness, V_g; thus we may write

$$\Delta_f \overline{w} = \mathrm{Cov}\,(w, g) / \overline{w} = V_g / \overline{w}, \tag{2.10}$$

where Δ_f emphasizes that this is a partial, Fisherian change, obtained by holding constant the contribution of each predictor.

Although Eq. (2.10) looks exactly like Fisher's fundamental theorem, I must add important qualifications in the following paragraphs. But first let us review the assumptions. The Price Equation is simply a matter of labeling entities from two sets in a corresponding way. The two sets are usually called parent and offspring. With proper labeling, the covariance and expectation terms follow immediately from the statistical definitions.

For any trait we can write $z = g + \delta$, where g is the sum of effects from a set of predictor variables, the effects obtained by minimizing the summed distances between prediction and observation (maximizing the use of information available from the predictors). This guarantees g is uncorrelated with δ. If we substitute into the Price Equation, the result in Eq. (2.7) follows immediately. Fisher was concerned with the part of the total change in fitness when the effect of each predictor is held constant, yielding Eq. (2.10). Thus Eq. (2.10) is obtained by using the best predictors of the trait substituted for the trait itself, and holding constant the effects of the predictors.

I close this section by reviewing a few technical details about Fisher's theorem. I discussed these issues extensively in Frank (1997e). Here I emphasize those points that will aid in the analysis of kin selection and Hamilton's rule.

From Eq. (2.10) and a bit of algebra given in Frank (1997e), the fundamental theorem can be expressed in terms of frequency change

$$\Delta_f \overline{w} = \mathrm{Cov}\,(w, g)\,/\,\overline{w} = V_g\,/\,\overline{w} = \sum (\Delta q_i)\,g_i,$$

where g_i is the breeding value of the ith element and, using definitions from the section on the Price Equation, the change in frequency of the ith element in the population caused by natural selection is

$$\Delta q_i = q_i' - q_i = q_i w_i\,/\,\overline{w} - q_i = q_i\,(w_i\,/\,\overline{w} - 1)\,.$$

This notation emphasizes Fisher's interpretation that natural selection directly causes changes in frequency, and only indirectly has consequences for changes in the effects of predictors via changes in breeding value. Thus the partial change caused by natural selection is the frequency change caused directly by natural selection, holding constant the effects of the predictors (breeding value).

I have given my equations for the fundamental theorem in terms of the frequencies of the aggregate elements, that is, the frequency of the aggregate i as q_i. In genetics the aggregate i would normally be an individual genotype, composed of a set of alleles (predictors) that comprise the genotype. In the notation of Eq. (2.5), the individual alleles are x_{ij}, for the number of copies of the jth allele in the ith individual. I denote frequencies for allele j as r_j. With this notation, I (Frank 1997e) showed the equivalence of the theorem expressed in terms of the aggregate elements or the individual predictors

$$\Delta_f \overline{w} = \sum (\Delta q_i) g_i = n \sum \left(\Delta r_j \right) b_j,$$

where b_j was defined above as the average effect (partial regression) for each allele and n is the maximum number of copies of an allele in each individual (ploidy).

This form shows that the partial change caused by natural selection, $\Delta_f \overline{w}$, is the frequency change of the predictor caused by selection, Δr_j, holding constant the effect of each predictor, b_j. Fisher (1958a) limited his discussion of the theorem to cases in which all frequency changes in the predictors (alleles) are caused directly by selection. Under this interpretation, the theorem holds only when selection is the sole force influencing frequency changes, and fails when mutation or other forces act on frequency. By contrast, I interpret the frequency change terms as partial changes caused by differential fitness. Under this interpretation the "partial frequency fundamental theorem" is universally true and provides a useful guideline for analysis of models such as kin selection (Frank 1997e).

Kin Selection

The next chapter is devoted to kin selection. But it is useful here to place the topic within the broader framework for the analysis of natural selection.

Hamilton's (1964a, 1970) famous rule provides a condition for the increase in altruistic characters

$$rB - C > 0,$$

where r is the kin selection coefficient of relatedness between actor and recipient, B is the reproductive benefit provided to the recipient by the actor's behavior, and C is the reproductive cost to the actor for providing benefits to the recipient.

We start our analysis, as before, by writing the character under study as $z_i = g_i + \delta_i$. For offspring derived from parental type i, $z_i' = g_i' + \delta_i'$. Because $\overline{\delta}' = \overline{\delta} = 0$, we have, as before, $\Delta\overline{z} = \Delta\overline{g}$, so we can work at the level of breeding values. Following Queller (1992a, 1992b) and the general approach of Lande and Arnold (1983), we begin with a regression equation for fitness

$$w = \beta_{wg \cdot G}\, g + \beta_{wG \cdot g}\, G + \epsilon,$$

where G is the average breeding value of the local group with which an individual interacts, $\beta_{wg \cdot G}$ is the partial regression of fitness on individual breeding value, holding group breeding value constant, $\beta_{wG \cdot g}$ is the partial regression of fitness on group breeding value, holding individual breeding value constant, and ϵ is the error term which, by least squares theory, is uncorrelated with g and G.

We can match this notation to standard models of kin selection (Queller 1992a, 1992b). The direct effect of an individual's breeding value on its own fitness, $\beta_{wg \cdot G}$, determines the reproductive cost of the phenotype. To match the convention that cost reduces fitness, we set $\beta_{wg \cdot G} = -C$. The direct effect of average breeding value in the local group on individual fitness, $\beta_{wG \cdot g}$, measures the benefit of the phenotype on the fitness of neighbors, thus $\beta_{wG \cdot g} = B$. The fitness regression can now be written as $w = -Cg + BG + \epsilon$. Substituting into the Price Equation, the condition for $\Delta\overline{z}$ to increase is equivalent to the condition for $\overline{w}\Delta\overline{g} > 0$, thus

$$\overline{w}\Delta\overline{g} = \mathrm{Cov}\,(w, g) + \mathrm{E}\,(w\Delta g)$$
$$= -C\mathrm{Cov}\,(g, g) + B\mathrm{Cov}\,(G, g) + \mathrm{E}\,(w\Delta g),$$

and, dividing by $\mathrm{Cov}(g, g) = V_g$, we obtain the condition for $\overline{w}\Delta\overline{g} > 0$ as (Frank 1997e)

$$rB - C > -\frac{\mathrm{E}\,(w\Delta g)}{V_g},$$

where $r = \mathrm{Cov}(G, g)/\mathrm{Cov}(g, g)$ is the kin selection coefficient of relatedness (reviewed by Seger 1981; Michod 1982; Queller 1992a).

This is an exact, total result for all conditions, using any predictors for breeding value. The predictors of phenotype may include alleles, group characteristics, environmental variables, cultural beliefs, and so on. If we use the Fisherian definition of partial change caused directly by natural selection, holding average effects constant, then the right side is zero and we recover the standard form of Hamilton's rule. Thus Hamilton's rule is an exact, partial result that applies to all selective systems, just as the partial frequency fundamental theorem is an exact, partial result with universal scope. Hamilton's rule may also be thought of as a kind of fundamental theorem, with the object of study a social character rather than fitness, and the causes of fitness separated between individual and social effects.

Several classical analyses of selection, such as Hamilton's rule and Fisher's fundamental theorem, assume constancy of average effects (see *Predictors and Additivity*, p. 18). Those statistical models share a point of view, from which one tends to overlook the details of how each particular combination of alleles (genotype) determines a particular character (phenotype).

2.3 Genotypes and Phenotypes

[A] system may be as broad or as narrow as we please depending upon the purpose at hand; and the data [parameters] of one system may be the variables of a wider system depending upon expediency. The fruitfulness of any theory will hinge upon the degree to which factors relevant to the particular investigation at hand are brought into sharp focus.

—Paul A. Samuelson, *Foundations of Economic Analysis*

To study the evolution of a phenotype, do we need to know its genetic basis? This is an important question, because I am headed for a simplified analysis that often ignores genetic details. Before arriving, it is useful to consider what is being left out. The main issue concerns how one chooses to separate factors into those fixed extrinsically (parameters) and those undetermined prior to analysis (variables).

Sex allocation provides a good illustration of the problem. I will discuss this topic fully in Chapter 9. Here I outline two contrasting models. The first emphasizes maximization of success in an economic analysis of phenotype. The second focuses on the dynamics of the hereditary particles (alleles) that determine phenotype.

PHENOTYPES AND MARKET SHARE

Any trait that increases its relative representation in the population will become common. The inexorable increase of successful traits is natural selection. This suggests that we could analyze how natural selection shapes traits by seeking those traits that maximize their relative success. In economic language, we seek traits that maximize market share.

For example, how does natural selection influence a parent's division of resources between sons and daughters? How many boys and how many girls? How much energy to devote to each? The economic problem is to divide a limited supply of resources into two distinct investment strategies, with the goal of maximizing market share relative to other competing families in the population. A model describing this investment problem is

$$w(x, y) = \frac{\mu(x)}{NE[\mu(x)]} + \frac{\phi(y)}{NE[\phi(y)]},$$

where $w(x, y)$ is the relative success of a mother as a function of her investment in sons, x, and daughters, y, subject to the constraint that $x + y = K$. A mother's ability to increase her representation through sons depends on the value of her sons, $\mu(x)$, relative to the total value of sons produced by families. This total is $NE[\mu(x)] = NE_\mu$, where N is the number of families and E_μ is the average value of sons in each family. Likewise, market share achieved through daughters is $\phi(y)$ compared with $NE[\phi(y)] = NE_\phi$.

The details of this model will be described in Chapter 9. Here I simply note that the split between sons and daughters, x and y, that maximizes relative success satisfies

$$\frac{\mu'(x)}{NE_\mu} = \frac{\phi'(y)}{NE_\phi}, \tag{2.11}$$

where the primes denote derivatives. The term $\mu'(x)$ is the marginal value of investment in sons. That return is standardized by the total value of sons, NE_μ. The right side is the marginal value of female investment standardized by the total value of female investment. This outcome follows the fundamental result of economic theory, the equilibration of marginal values. The factors in the denominator transform the problem into maximization of market share.

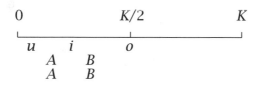

Figure 2.1   Three types of dominance relations for a single diploid locus. The phenotype space in this problem is a number on the continuous interval $[0, K]$. The phenotypes of the homozygotes, AA and BB, are shown. The heterozygote, AB, may have a phenotype smaller than both homozygotes, labeled as u, for underdominance. If the heterozygote is between the homozygotes, it is labeled i, for intermediate dominance. If the heterozygote is larger than both homozygotes, it is labeled o, for overdominance.

If returns are linear for each sex, $\mu(x) = ax$ and $\phi(y) = by$, with a and b arbitrary, positive constants, then Eq. (2.11) reduces to

$$\frac{a}{Na\bar{x}} = \frac{b}{Nb\bar{y}}.$$

Thus

$$\bar{x}^* = \bar{y}^*, \tag{2.12}$$

where $*$ denotes equilibrium. Under linear returns, an equal split between investment in males and females is favored at the population level. This result, first given by Fisher (1958a, 158–160), plays an important role in the foundations of social evolution. However, this equal allocation theory has been overused because the required restrictions on $\mu(x)$ and $\phi(y)$ are often forgotten (see Chapter 9).

Genetics: Constraints on Paths of Phenotypic Evolution

The phenotypic model hides many details. For example, if all genotypes in the population produce family allocations that underweight sons, $x < K/2$, then the population allocation is $\bar{x} < K/2$. Equal allocation may be favored, but a phenotype cannot evolve if no genotype produces that phenotype. Even if a genotype that produced $x = K/2$ existed, the population might be stuck at an alternative equilibrium.

Fig. 2.1 shows various assumptions about the relationship between genotype and phenotype. Sex allocation in this example is controlled by a pair of alleles, one inherited from the mother and one from the father (a single diploid locus). The sex allocation, expressed as resources invested

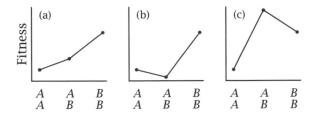

Figure 2.2   The three cases of dominance in Fig. 2.1 shown on a fitness scale. From left to right, intermediate dominance, underdominance, and overdominance.

in sons, x, ranges from zero to K. Suppose initially that the population has only the A allele: everyone is an AA homozygote with a phenotype $x < K/2$. Then one of the A alleles mutates into a B allele with a different phenotypic effect. The BB homozygote has a phenotype x closer to $K/2$, but still less than this midpoint.

Does this rare B allele spread in a population fixed for the A allele? There are three cases:

1. The AB heterozygotes have a phenotype x intermediate between AA and BB. The i in Fig. 2.1 shows the location of the AB phenotype. Since $K/2$ is the favored phenotype, and i is closer to $K/2$ than the common AA genotype but farther than the BB genotype, AB exhibits intermediate dominance on the fitness scale. This is shown in Fig. 2.2a.

Initially, the rare B allele will exist in AB genotypes because one B will very rarely meet another B. Since AB has a higher fitness than AA, selection carries the B to higher frequency. The BB genotype has a higher fitness than AB, so selection continues to push the frequency of B higher until everyone is BB and the A allele has been eliminated. The mutation B has shifted the population closer to the optimum of $K/2$.

2. The AB heterozygotes have a phenotype x smaller than AA. The u in Fig. 2.1 shows the location of the AB phenotype. Since a larger value is the favored phenotype, and u is smaller than either AA or BB, the AB genotype is underdominant on the fitness scale. This is shown in Fig. 2.2b.

The rare AB types that occur after the B mutation arises have lower fitness than the common AA type. Thus the frequency of B declines until the B allele disappears from the population. This extinction occurs in spite of the fact that the BB homozygote has a higher fitness than the resident AA type. The improved BB equilibrium cannot be reached when AA is common and AB is underdominant.

3. The AB heterozygotes have a phenotype x greater than both AA and BB. The o in Fig. 2.1 shows the location of AB. In this case the heterozygote is closer to the favored phenotype than either homozygote, and is called overdominant. This is shown in Fig. 2.2c.

The rare AB genotypes that occur after after the B mutation arises have higher fitness than the common AA type. Thus the frequency of B initially increases, and the population contains a mixture of the two pure genotypes and the mixed heterozygote. The heterozygote has the highest fitness, but the population cannot become purely heterozygous because, in each generation, an individual inherits one allele from each parent. Some individuals will, by chance, get two A alleles, others will get two B alleles, and yet others will get the favored mixture of alleles. This mixed condition of A and B alleles stabilizes at an equilibrium, polymorphic state.

These three cases only hint at the potential dynamic complexities of genetics. They do, however, show that economic maximization of fitness (market share) can easily be prevented by the way in which phenotypes are specified by the hereditary mechanism.

Resolution: The Spectrum of Mutations

The genetic models assume the range of genetic variability to be given by fixed parameters. The phenotypic models seemingly ignore genetics altogether. Since the genetic models show that phenotypic maximization is not a necessary outcome of selection, how can one justify using the simpler, phenotypic models?

Suppose that we also consider the range of mutations that are possible and how frequently they occur. The genetic assumptions are now variables that change rather than fixed parameters. We have pushed back the level of explanation, and now take the origin of genetic variation as the controlling parameter.

If, for example, the population is fixed at AA, and an underdominant mutation, B, occurs, as in Fig. 2.2b, the B allele does not increase. Underdominance prevents the population from moving closer to the predicted phenotypic optimum. However, the next mutation to come along, C, may have intermediate dominance, allowing individuals to move closer to the optimum. If a sufficient diversity of mutations is allowed, with varying dominance and magnitude of effect, then eventually the population will converge on the maximum. Once there, no mutation will displace it. Thus, genetics determines the rate of transitions, but the final stop is independent of genetics (Hammerstein 1996).

If one is concerned with short-term responses to selection pressures, then explicit genetic theory and matching observation would be valuable (Eshel 1996). This has been difficult because the genetics of interesting behavioral traits are rarely known.

Extant genetics is less important than the spectrum of mutations over long periods of time. Because mutations are rare events, it is very difficult to obtain observations that would aid in predicting the time course of evolutionary change. These theoretical and practical reasons suggest that the phenotypic approach is the only viable method for study of social evolution (Grafen 1991).

Theory with explicit genetics and assumptions about mutation can be useful. Such models allow one to quantify how often and by how much a simplified phenotypic model differs from models with restricted assumptions about genetics and mutation. However, theoreticians devoted to this subject have not concerned themselves with this practical question, probably because it can be studied only by approximate computer methods rather than by the quasi-physical dynamics and theorems that this research group prefers (see Gayley and Michod 1990, for an interesting exception).

Some limits must be placed on possible phenotypes. For example, a mutation that caused better performance in every dimension would, of course, increase in frequency. All useful theories must impose constraints on the phenotypic space. The source of such constraints may arise from genetics, physics, development, and so on. Plausible constraints are constructed from prior data or by hypothesis. This issue has been summarized by Parker and Maynard Smith (1990).

2.4 Comparative Statics and Dynamics

> Often in the writings of economists the words "dynamic" and
> "static" are used as nothing more than synonyms for good
> and bad, realistic and unrealistic, simple and complex. We
> damn another man's theory by terming it static, and adver-
> tise our own by calling it dynamic. Examples of this are too
> plentiful to require citation.
>
> —Paul A. Samuelson, *Foundations of Economic Analysis*

I have two major goals in this book. First, I extend the classical statistical
models of social evolution described in the previous sections. Second,
I develop new analytical methods within the framework of comparative
statics. This section provides a brief introduction to comparative statics.
The following section outlines the new analytical methods.

THE IMPORTANCE OF COMPARISON

Fisher's sex allocation theory predicts equal investment of parental resources in sons and daughters (Eq. (2.12)). How can such a theory be
tested? One common approach is to estimate the resources invested in
each sex and compare the fit to the predicted equal division. There are
several problems with fitting. A fit requires a precise estimate for investment, for which there is no clear and universally applicable working
definition. The prediction of equal allocation requires a strict assumption about the functional forms of returns on investment in each sex
(e.g., $\mu(z) \equiv \phi(z)$). Further, one cannot exclude alternative theories
that, for some parameter values, also predict equal allocation. A fit provides a sample size of one to test a particular theory versus alternative
causal explanations. Finally, lack of fit provides limited information
about what aspect of the theory requires further study.

Comparison solves some of these problems. For example, let returns
on male investment be $\mu(z) = z^s$ and returns on female investment be
$\phi(z) = z^t$, where $0 < s, t \le 1$. If all families have the same resource
level then, from Eq. (2.11), the equilibrium allocation ratio of males to
females in a population is $c : 1$, where $c - s/t$.

We now have a simple comparative prediction: as c rises, the relative
investment in males is expected to increase. A precise measure of c is
impossible. But, in comparison among cases, it may be easy to determine

how c changes. For example, in one case it may be that returns on male investment diminish faster than returns on female investment, $c < 1$, whereas in a second case the reverse is true. The theory predicts a switch from female-biased investment ($c < 1$) to male-biased investment ($c > 1$). If observations fail to match the theory, we can reject an equilibrium controlled by c as an explanation for sex allocation.

This example illustrates the fundamental role of comparison in the formulation of a theory. I will develop the subject of sex allocation in Chapter 9.

Dynamic Assumptions in Comparative Statics

A system at equilibrium does not change. Thus equilibrium is often referred to as a static condition. Comparison among predicted equilibria as a function of a parameter, as in the sex allocation example with parameter c, is called comparative statics (e.g., Schumpeter 1954; Samuelson 1983).

Comparative statics requires that populations change more quickly than parameters. If the parameter c varied rapidly but populations adjusted only slowly to those changes, then an observed population would probably not be close to an equilibrium for the current value of c.

Comparative statics may mislead if disequilibrium is sufficiently widespread. The arguments for pushing ahead with comparative statics are mainly practical rather than formal:

1. A hypothesis of disequilibrium is, by itself, irrefutable. A causal model can take on almost any state when the causes of disequilibrium are not specified.

2. Dynamics are interesting only when predictions can be formulated in a comparative way. How do observable dynamics change as a measurable parameter changes? Theoretical complexity and the lack of suitable data put comparative dynamics out of reach for most subjects.

3. A practical defense of comparative statics requires only usefulness, rather than a formal guarantee of success. Practically, one requires that directional tendencies predicted by comparative statics are dynamically valid often enough that, on average, something is learned. When a particular prediction fails, one cannot separate the

approximate nature of the theory from the possibility that the explanation is incorrect. Only across many cases to which the theory may apply can confidence be improved.

These problems of inference can often be studied in a formal way by computer analysis. One constructs a dynamical model of evolutionary change, complete with a specific spectrum of mutational effects. Then, one builds an evolving biological system in the computer; the program measures only those attributes that an experimenter could actually measure. Those data are analyzed, and the inferences are compared with the true evolutionary trend in the evolving computer population. Without such an analysis, it is often impossible to determine the power of a particular sampling scheme for discriminating among competing explanations.

2.5 Maximization and Measures of Value

An engineer finds among mammals and birds really marvelous achievements in his craft, but the vascular system of the higher plants, which we do not understand, has apparently made no considerable progress. Is it like a First Law, not a great engineering achievement, but better than anything else *for the price?* Are the plants not perhaps the real adherents of the doctrine of marginal utility, which seems to be too subtle for man to live up to?
—R. A. Fisher, Letter to Leonard Darwin

The job of doing comparative statics is much easier when one can use maximization techniques to search for local equilibria. If the desired result is a maximum, then the problem reduces to three steps. First, specify the appropriate value function to be maximized. Second, describe constraints on the variables. Third, use the standard tools of calculus to find local maxima subject to the constraints.

Relative reproductive rate, or fitness, is the measure of value one uses to study the consequences of natural selection. Many fundamental insights about natural selection concern the proper formulation of a fitness function for use in maximization methods. Some examples follow.

REPRODUCTIVE VALUE

If an allele can produce an effect (trait) that increases its future frequency, then the associated trait will increase in prevalence. A proper analysis of selection projects future consequences for the present distribution of traits. The allele with the greatest rate of increase will determine biological pattern in the future.

Analysis of the future entails prediction. Most biological theory is, however, concerned with explanation of the past. A mathematical statement about traits that maximize projection into the future provides hypotheses about how past selection has shaped the current distribution of traits.

How does one measure the reproductive consequences of a trait? That depends on the trait. For the design of vascular structure in plants, the natural measure is a simple count of the number of successful offspring. Alleles that influence vascular design will spread or disappear according to the number of successful offspring produced by the plants in which the alleles live.

Suppose the trait is the distribution of parental resources to offspring of different ages. Let our organism live n years. The number of offspring produced, a, is the same in each year. The probability of survival to the next year is p, until certain death after the nth year (start of the $n + 1$st year).

The expected future contribution of each offspring depends on its age, x. In the current year it will produce a offspring, it will survive with probability p to produce a offspring in the next year, and so on. Thus reproductive value, $v(x)$, is

$$v(x) = a \sum_{i=0}^{n-x} p^i = a \left(\frac{1 - p^{n-x+1}}{1 - p} \right),$$

where the right side of the equation is produced by the standard identity for geometric series.

How should a parent distribute limited resources among offspring of different ages? This is a common sort of question, which is really a shorthand for the following. Suppose there is genetic variation that influences a parent's behavior with regard to distribution of resources to offspring. Which genotypes will be favored? How will natural selection shape parental behavior?

We must search for allelic effects that maximize reproductive rate. Older offspring have a lower future expectation of reproduction and therefore may be less valuable than younger offspring. Consider two cases. First, suppose that parental resources influence the survival of offspring to the following year. Then offspring reproductive value is

$$v(x, \delta_x) = a + a(p + f(\delta_x)) \sum_{i=0}^{n-x-1} p^i \qquad x = 0 \ldots n-1$$

$$= a \qquad\qquad\qquad x = n$$

where $f(\delta_x)$ is the effect on offspring survival to the next year given an additional input of parental resources of δ_x. The parent's problem is to divide its limited resources among offspring of different ages. If f is a diminishing-returns function, then by the theory of marginal values the maximum occurs when

$$\frac{\partial v}{\partial \delta_x} = \frac{\partial v}{\partial \delta_y} \qquad x, y = 0 \ldots n-1$$

and $\delta_n = 0$ because offspring of age n die in the following year. From this condition the equilibrium must satisfy

$$f'(\delta_x) = \frac{K}{\sum_{i=0}^{n-x-1} p^i} = \frac{K(1-p)}{1-p^{n-x}} \qquad x = 0 \ldots n-1,$$

where K is a constant determined by the amount of parental resources available for distribution. The right side of the equation increases in x, so older individuals must be associated with higher marginal survival, $f'(\delta_x)$. Higher marginal survival occurs with lower values of δ_x. Thus parents are favored if they allocate fewer resources to relatively older offspring.

In this first case, parents influence offspring survival for one year, from the current year to the following year. The second case assumes that parents influence offspring reproduction for one year, the current year. In this model the effect of parental investment δ_x to offspring of age x is

$$v(x, \delta_x) = a + g(\delta_x) + a \sum_{i=1}^{n-x} p^i,$$

where $g(\delta_x)$ is the effect of parental investment on reproduction in the current year. Since $\partial v / \partial \delta_x$ is independent of age, x, parents are favored if they treat offspring of all ages equally.

In the first model, changes in offspring survival through the current year provide marginal returns in proportion to future reproduction. The favored distribution of resources is age-dependent because young offspring have a higher future expectation of reproduction than old offspring. In other words, young offspring have higher reproductive value than old offspring.

In the second model, parental aid of offspring reproduction helps all offspring equally in the current year, and has no consequences for future reproduction. Thus parents are favored if they distribute resources independently of age.

Reproductive value is a method of weighting individual values so that simple maximization techniques can be used. I will discuss in Chapter 8 various demographic and genetic factors that influence reproductive value.

KIN SELECTION

In the previous section a trait influenced its future frequency by direct effects on offspring. The value of investment in each offspring was measured by marginal change in reproductive value, that is, by marginal change in expected contribution to the future gene pool.

The success of a trait may also be affected by social partners with correlated traits. I previously analyzed social interaction by partitioning the fitness consequences of a trait into individual and social components (see *Kin Selection*, p. 23). Here I briefly extend the analysis to show that kin selection coefficients have a broader interpretation as measures of value.

HAMILTON'S RULE

An individual's fitness, w, can be written as a function of its own phenotype, y, and its neighbors' average phenotype, z,

$$w (y, z) = \beta_{wy \cdot z} y + \beta_{wz \cdot y} z + \epsilon,$$

where the β's are partial regression coefficients and ϵ is uncorrelated with y and z (Queller 1992a, 1992b). The effect of y on w, holding z constant, is the effect of an individual's phenotype to its own fitness, so we may say that the cost of an individual's phenotype on its fitness is $C = -\beta_{wy \cdot z}$. Similarly, the effect of z on w, holding y constant, is the effect of the neighbors' phenotype to our focal individual's fitness. Thus we can call the benefit of the neighbors on our focal individual $B = \beta_{wy \cdot z}$. Substituting, we have

$$w(y, z) = -Cy + Bz + \epsilon$$

My goal is to study the evolution of the allelic effect, x, via its phenotypic effects on an individual and its neighbors. From the Price Equation (2.3), if one holds constant the average effect of the allele over time, $\Delta x = 0$, then the condition for an increase in $\bar{x}$ is $\mathrm{Cov}(w, x) > 0$. Thus the condition for increase is

$$\mathrm{Cov}(w, x) = -C\,\mathrm{Cov}(y, x) + B\,\mathrm{Cov}(z, x) > 0,$$

under the assumption that $\mathrm{Cov}(\epsilon, x) = 0$, that is, the allele x influences fitness only through its effect on the phenotypes y and z. Dividing by $\mathrm{Cov}(y, x)$, we recover Hamilton's rule, $rB - C > 0$, where the definition of relatedness, r, between individual and neighbor is

$$r = \frac{\mathrm{Cov}(z, x)}{\mathrm{Cov}(y, x)}. \tag{2.13}$$

RECOVERY OF MAXIMIZATION

Hamilton's rule provides a measure of valuation, r, for comparing social components of fitness. But the rule itself is given as an inequality. It would be useful to express valuation in a way that allows us to use maximization methods. Such methods provide powerful tools for developing comparative statics.

Taylor and Frank (1996) showed how to incorporate kin selection into standard optimization methods. To continue with the above example, with fitness function $w(y, z)$, individual phenotype y, neighborhood phenotype z, and allelic value x, suppose we differentiate w with respect to x. Using the chain rule, we obtain

$$\frac{dw}{dx} = \frac{\partial w}{\partial y}\frac{dy}{dx} + \frac{\partial w}{\partial z}\frac{dz}{dx}.$$

In the typical maximization analysis we would evaluate $dw/dx = 0$ at $x = x^*$. The technical problem here is what to make of the derivatives with respect to y and z, because relations to x may be statistical rather than functional. However, dy/dx and dz/dx are slopes of phenotype on allelic value, and if we replace these by the corresponding statistical regression coefficients we get

$$\frac{dw}{dx} = \frac{\partial w}{\partial y}\beta_{yx} + \frac{\partial w}{\partial z}\beta_{zx}.$$

Dividing by β_{yx} we get the rate of change in fitness with changes in individual phenotype, y, as

$$\Delta w = \frac{\partial w}{\partial y} + r\frac{\partial w}{\partial z}$$

$$= -C_m + rB_m,$$

where r, given in Eq. (2.13), is relatedness of an individual, y, to a random neighbor, z. From the previous section, it is clear that $-\partial w/\partial y = C_m$ is the marginal cost of an individual's behavior to its own reproduction, and $\partial w/\partial z = B_m$ is the marginal benefit of the behavior to a neighbor.

This marginal version of Hamilton's rule shows that an equilibrium, $\Delta w = 0$, often satisfies the condition that marginal costs and benefits be equal, $C_m = rB_m$. Note how the relatedness coefficient, r, plays the role of currency translation between marginal effects on direct and indirect reproduction.

I will rework the theory of kin selection in later chapters, including these brief examples as special cases. The important point here is that the standard maximization method can be adapted to analyze interactions if we simply replace derivatives by the appropriate regression coefficients. The method can also be extended to handle social interactions when different neighbors have different reproductive values. This extended method provides a simple, general maximization approach for the study of social interactions.

RELATEDNESS, GENEALOGY AND INFORMATION

The common interpretation of relatedness is genealogical kinship. Relatives share similar genes by common ancestry. An allele may increase its frequency by enhancing the reproduction of the body in which it lives or by increasing the reproduction of the same allele in a different body. In this context, the regression coefficient of relatedness, r, measures value by weighing the excess number of copies of itself in different bodies when compared to the population average.

Genealogical kinship is probably the most important factor in social evolution. But the following example shows that kin selection must be a subset of a broader phenomenon.

Let pairs of individuals interact in a social situation. Label the individuals of a pair as player I and player II for convenience. The average probability that player I acts cooperatively is $\bar{p}$, and the average probability that player II acts cooperatively is $\bar{q}$. A particular allele causes an increase in player I's probability of cooperating by an amount x, so that $p = \bar{p} + x$. This increase in the tendency to cooperate reduces player I's own reproduction by an amount Cx.

We can, by standard regression, always write the tendency of player I's partners to cooperate as

$$q - \bar{q} = r(p - \bar{p}) + \epsilon$$
$$= rx + \epsilon,$$

(2.14)

where the expected value of ϵ is zero by regression theory. The regression coefficient, r, is the slope of deviations in partner phenotype, $q - \bar{q}$, relative to deviations in the actor's genotype, x. The expected deviation in the partner's phenotype is rx, and that change in cooperative behavior has a benefit to the actor of rxB.

The condition for x to increase is that benefit be greater than cost, $rxB - Cx > 0$, which is equivalent to

$$rB - C > 0.$$

This is, of course, Hamilton's rule. The point here concerns the interpretation of the coefficient, r. This coefficient is simply the regression of partner phenotype on actor genotype. Nothing in the derivation suggested that the partners must be related by common kinship. Indeed, the partners could be different species, and the same derivation would apply. Given that kinship is not required, what does the regression coefficient mean?

Regression coefficients predict the value of one variable based on the given value of a second, predictor variable. Put another way, regression coefficients describe conditional information: given the predictor, a more accurate estimate of the outcome is possible. In this case, the predictor is the genotype of an actor. The phenotypes of partners are the outcome variables that are estimated with improved accuracy.

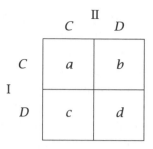

Figure 2.3   The entries show the payoff to player I given strategies by two players in an encounter. The alternative strategies for each player are C and D, for Cooperate and Defect.

Hamilton's rule describes how selection favors maximal use of the information available. Genealogical kinship is one form of information that is commonly available. But other types of information influence selection in the same way, as illustrated by the example in which partners are different species. I discuss relatedness as a coefficient of information in Chapter 6.

Game Theory, ESS

Much of the recent literature on social behavior is concerned with game theory. The popularity of this approach arose from Maynard Smith and Price's (1973) paper, which outlined the conditions for an Evolutionarily Stable Strategy (ESS). Briefly, a strategy (phenotype) is evolutionarily stable if all members of a population adopt the strategy and any rare deviant individual has lower fitness than the normal types. This is, of course, a simple criterion for a local equilibrium, and we have already used it in previous sections. The ESS also turns out to be closely allied to the classic Nash equilibrium of formal game theory (Maynard Smith 1982).

MONOMORPHIC EQUILIBRIUM WITH CONTINUOUS PHENOTYPES

The history of game theory and its connections to other fields are interesting, but sometimes obscure how simple the methods really are in practice. I illustrate this with the game shown in Fig. 2.3. Phenotypes are continuous in this case. Thus we can write a function to describe individual fitness and then maximize the function with respect to small variants about a potential equilibrium. Let p be the probability that our focal individual, player I, plays strategy C. The frequency of opponents, player II, that play strategy C is q. Individual fitness, w, is

$$w(p,q) = pqa + p(1-q)b + (1-p)qc + (1-p)(1-q)d, \quad (2.15)$$

where, in the first term, our focal individual plays C with probability p, its opponent also plays C with probability q, and the payoff for this combination is a. Similar descriptions apply to the other three combinations.

By the methods given above, let x be a small deviation in the average effect of an allele influencing phenotype. Evaluating $dw/dx = 0$, replacing the phenotypic derivative with r, and evaluating at $p = q = p^*$ yields

$$p^* = \frac{b - d + r(c - d)}{(1 + r)(b + c - a - d)},$$

which has been obtained previously by other methods (see Chapter 5). However, this differentiation method is perhaps the simplest, and maintains a consistent maximization approach for various types of problems. Other approaches appeal to ad hoc methods for kin selection and other special circumstances.

A mixed equilibrium, $0 < p^* < 1$, always obeys an important economic principle. When individuals can vary their allocations to different strategies by small amounts, then a mixed equilibrium always occurs when the marginal fitness returns are equal for the different strategies. The marginal returns are measured with respect to the fitness of allelic variants of small average effect.

EQUILIBRATION OF FITNESSES FOR TWO ALLELES

The previous model assumed the potential for small variations in the phenotype, x. Another common problem assumes that two competing genotypes have very different phenotypes. What is the equilibrium frequency of the two types in the absence of mutations to other phenotypes?

This question raises an important point about natural selection. We have, thus far, been obtaining equilibrium points by maximization, but an equilibrium under natural selection guarantees only the weaker condition that fitnesses of all alleles be the same. If the fitness of some allele is greater than the others, it will increase in frequency until fitnesses are either equilibrated or the competing alleles are driven to extinction.

Maximization was appropriate in the prior cases because we allowed small variants near a candidate equilibrium, and only came to rest when those variants had lower fitness. Thus the equilibrium is a local maximum. In the present case there are no variant phenotypes, only two fixed types, and so we can use only the principle of equilibration to obtain an equilibrium condition.

The standard game theory model with discrete phenotypes assumes "pure" strategies, in which each player always plays either C or D (Maynard Smith 1982). Thus $p = 1$ for always play C, and $p = 0$ for always play D. Mixtures of the two types are possible in the population. The goal of the analysis is to find the equilibrium mixture, $\bar{p}^*$.

There are various approaches to studying interactions between kin or, more generally, between correlated players (Chapter 5). The most general approach to correlation between players is given in Eq. (2.14), where the expected value of player II's strategy, q, given player I's strategy, p is

$$E(q - \bar{q}|p) = r(p - \bar{p}).$$

I assume that both players come from the same population; thus $\bar{p} = \bar{q}$. The expected value of q given p is therefore

$$q = \bar{p} + r(p - \bar{p})$$

$$= (1 - r)\bar{p} \qquad\qquad p = 0$$

$$= r + (1 - r)\bar{p} \qquad\qquad p = 1.$$

Fitnesses for haploid individuals can be obtained from Eq. (2.15). Individuals always playing strategy C, with $p = 1$, have fitness $w_C = w(0, q)$, and those playing strategy D, with $p = 0$, have fitness $w_D = w(1, q)$. Substituting for q and expanding these definitions yields

$$w_C = ra + (1 - r)[\bar{p}(a - b) + b]$$

$$w_D = d + (1 - r)\bar{p}(c - d).$$

The solution for $\bar{p} = p^*$, obtained from $w_C = w_D$, is

$$p^* = \frac{b - d + r(a - b)}{(1 - r)(b + c - a - d)},$$

suggesting that the constrained phenotypes have equilibrated fitnesses at this point, but individual fitnesses are not at a local maximum with respect to variations in the probability that a particular individual will play either strategy C or strategy D.

DIFFICULTIES

Maximization works only if small variants are present, there is a dynamic path to the maximum, and the maximum is stable. For discontinuous phenotypes, equilibration of fitnesses also requires stability. Against this, many types of feedback destabilize, and nonlinearities create multiple equilibria that may block the pathway to a particular local maximum. Much of the theoretical literature is devoted to these problems. Their importance in practice is unknown and perhaps unknowable—hence the enduring value of simplified comparative statics.

Another issue concerns the maximization of value when there are stochastic fluctuations in payoff. The theories given here and below assume that stochastic factors and risk can be ignored. Theories of value with uncertainty are well developed in economics and have been applied to behavioral and genetic models in biology (e.g., Gillespie 1977; Real 1980; Tuljapurkar 1990).

Three special features of biology differentiate the theory of value under risk from standard concepts in economics (Frank and Slatkin 1990b). First, the evolutionarily relevant measure of value is relative success, or market share. This requires that allelic success be divided by average success of the population. This ratio of two correlated random variables introduces some interesting complications into standard economic theories of risk. Second, each individual faces uncertain returns in the accumulation of resources and the production of offspring. This individual variability has evolutionary consequences to the extent that such phenotypic variation influences genetic change. This leads to the third issue. Each allele occurs in many different individuals. Evolutionary consequences depend on the average success of an allele across its individual instances.

Difficulties can be studied. For dynamic complications, one needs to know how often simplified economic analyses will mislead. This depends on the particular hypothesis under study. One approach is to model evolving populations in the computer. Samples from those artificial populations provide information about how often similar sampling schemes in natural populations would cause incorrect inference. This approach is sometimes used in population genetics (e.g., Frank 1996c), but I do not know of cases in which this method has been applied to social evolution. Gayley and Michod (1990) used a computer model to study the dynamics of social evolution, but they did not consider the problems of sampling natural populations with respect to testing particular comparative hypotheses.

The problems of uncertainty have been well studied for resource acquisition (Stephens and Krebs 1986) and life history (Tuljapurkar 1990). Only a few authors have considered topics that are more directly social (e.g., Frank 1990b; McNamara 1995). On the whole, dynamic complications and uncertainty have not been developed in a useful way for social problems, and will not be discussed in the remainder of the book.

3 Hamilton's Rule

> Considerations of genetical kinship can give a statistical re-
> association of the [fitness] effects with the individuals that
> cause them.
>
> —W. D. Hamilton, "Selfish and spiteful
> behaviour in an evolutionary model"

Many traits decrease the reproductive success of their bearer and raise
the success of particular neighbors. The most striking case is sterility
of worker castes in social insects (Wilson 1971). The workers forgo their
own reproduction and devote their lives to raising the offspring of one
or a few royal members of the colony.

Less spectacular cases pervade nearly every aspect of ecology and
behavior. For example, some "prudent" parasites appear to use host resources more slowly than the rate that would optimize their reproductive success. The outcome is lower relative success for the individual,
but a higher level of productivity for the group of parasites in a host
(Herre 1993).

Most students of natural selection, prior to Hamilton's (1964a, 1964b)
work, explained traits only by their direct effect on individual reproduction (see Hamilton 1972, for discussion of the history). Insect sterility
and prudent parasites are puzzling from this individual view. These
traits lower the reproduction of their bearers, and therefore cannot be
explained purely by their direct effect on individuals.

Hamilton emphasized that traits change in frequency by two routes.
First, a trait may influence the reproduction of the individual that expresses the trait. Second, a trait may influence the reproduction of
neighboring individuals that possess a nonrandom sample of genes for
the trait.

Hamilton was particularly interested in "altruistic" traits that reduce
the reproduction of the actor but enhance the reproduction of neighbors. The net rate of increase of genes causing such a trait depends on
the reduction in individual reproduction compared with the increase in
neighbor reproduction. Reproduction by neighbors must be weighted

by the neighbors' deviation in frequency from the population average. If neighbors are closely related, then there is a higher probability than average that they share similar genes, and altruistic behaviors may be frequent. If neighbors are distant relatives, then direct reproduction is the only pathway by which genes may increase their own frequency, and altruistic traits are expected to be rare.

Hamilton (1964a, 1964b, 1970) introduced a mathematical theory to study the evolution of social traits. He expanded the theory of natural selection from one that focused solely on individual reproduction to a theory that could analyze the combined effects of individual reproduction and reproduction by relatives. Hamilton summarized his work in an elegant result, now known as Hamilton's rule. Subsequent debates have focused on the interpretation and validity of this rule.

3.1 Overview

Hamilton's rule states that a behavior increases in frequency when

$$rB - C > 0,$$

where r is the kin selection coefficient of relatedness between an actor and recipient of a particular behavior, B is the reproductive benefit conferred on the recipient, and C is the cost to the actor in direct reproduction. Hamilton (1970, 1972) emphasized that the rule depends on several assumptions, including weak selection, additivity of costs and benefits of fitness components, and a special definition of relatedness that uses statistical correlations among individuals rather than genealogy to describe similarity. The full mathematical details have been reviewed by Seger (1981), Michod (1982), Grafen (1985), and Queller (1992a, 1992b).

There are, roughly speaking, two schools of thought on Hamilton's rule. On one side, the limitations have been exposed formally by a wide variety of models in which Hamilton's rule fails (e.g., Charlesworth 1978; Uyenoyama and Feldman 1982; Karlin and Matessi 1983). On the other side, Michod (1982), Grafen (1985), and Queller (1992a, 1992b) have supported Hamilton's rule as a fundamental evolutionary principle by categorizing exceptions to the rule under the few well-understood headings that Hamilton pointed out in his original work.

The size and complexity of the literature has led to some confusion. Can one use simple reasoning based on kin selection to understand the evolution of complex behaviors? Is Hamilton's rule the proper simplification? Or, are the exceptions to simplified methods so numerous that one must formulate a complex genetic model to understand each social trait and its special conditions?

I take a middle position between these opposing views. On the one hand, kin selection, when properly used, is a powerful analytical tool. Complex problems can be reduced to simple models in which the biological interactions are clarified. On the other hand, the common practice of applying $rB - C > 0$ to reason about social evolution often fails. It is not so much a failure of the rule, but that the rule as stated hides too much. The inviting simplicity leads to hasty conclusions without careful specification of the biological interactions and the control of phenotypes. My view is that overly simplified analyses based on kin selection are too common, but that the full power of kin selection as an analytical tool is rarely employed.

Given the complexity of the subject, it is useful to follow the history of some technical issues. I begin with Hamilton's (1970) derivation. I then turn to Queller's (1992a, 1992b) elegant quantitative formulation, which uses the statistical approach outlined in Chapter 2. Queller's work places Hamilton's rule in the broader context of multivariate selection and causal analysis (Lande and Arnold 1983).

The prior work by Hamilton and Queller leaves several conceptual issues unresolved, and fails to provide practical guidelines for the solution of common problems. I extend the concepts and methods of kin selection in the following chapter. Later chapters illustrate the power of these extended methods by solving a wide array of problems in social evolution.

3.2 Hamilton's 1970 Proof

In the late 1960s George Price explained his covariance formulation of natural selection to W. D. Hamilton (1996, 171–176). Hamilton quickly realized that this covariance method provided a new way to study social behavior. This led to Hamilton's (1970) publication, which I summarize in this section (see Grafen 1985, for further details). I use Hamilton's notation in this section to allow comparison with the original publication. My notation in other sections matches the style of this book, which differs from Hamilton's.

DIRECT FITNESS

Hamilton first wrote an expression for the fitness of an individual as affected by its own phenotype and the phenotype of its neighbors. This simple description of individual fitness is called *direct* fitness, or sometimes *personal* or *neighbor-modulated* fitness.

The direct fitness of individual j in a population is

$$w_j = 1 + \sum_i s_{ij},$$

where s_{ij} is the effect of the ith social partner on the fitness of the jth individual. The term s_{jj} is the effect of the individual on itself.

I follow the frequency of an allele, A, which occurs at a diploid locus (one allele from mother and one allele from father for each gene of an individual). The allele frequency of A in each individual, j, is $q_j = (0, \frac{1}{2}, 1)$. I use the Price Equation to calculate the change in allele frequency. Repeating Price's formula from Eq. (2.3),

$$\overline{w}\Delta\overline{q} = \mathrm{Cov}\,(w, q) + \mathrm{E}\,(w\Delta q).$$

The term Δq describes the change in allele frequency transmitted by an individual. I make the standard population genetic assumption that the allele frequency transmitted by an individual, in its successful gametes, is the same as q_j, the allele frequency in the adult. Thus, if $q_j = 1/2$, then the adult transmits the allele A to one-half of its offspring. This standard assumption about transmission implies that $\Delta q = 0$. Thus the change in allele frequency is determined entirely by the covariance term

$$\overline{w}\Delta\overline{q} = \mathrm{Cov}\,(w, q) = \frac{1}{n} \sum_j \left(\sum_i s_{ij} \right) \left(q_j - \overline{q} \right). \tag{3.1}$$

This form matches the social effects, s, with the individual affected. The total of all social effects, $\sum_i s_{ij}$, determines the direct fitness of the jth individual. But it is difficult to predict how natural selection will affect the social characters, s, and the direct fitness. Each individual is influenced by a diversity of social partners, and each social partner may have differing reproductive interests.

INCLUSIVE FITNESS

Hamilton showed that social behavior may be easier to interpret if we take the point of view of the individuals who control behavior (actors) rather than those who receive the effects (recipients). Reassociation of the fitness effects in Eq. (3.1) can be achieved as follows. Express the allele frequency in the jth individual as a function of the allele frequency in the ith individual, q_i, yielding

$$q_j - \bar{q} = b_{ij} (q_i - \bar{q}) + \epsilon_j.$$

I have made a few minor changes from Hamilton's definitions. I define the b's as deviations that can be positive or negative, such that the total deviations are necessarily zero, $\sum_j b_{ij} = 0$. The ϵ_j are the unexplained part of the deviation, with the ϵ's summing to zero. Substitution into Eq. (3.1) yields

$$\bar{w}\Delta\bar{q} = \frac{1}{n} \sum_i x_i (q_i - \bar{q}), \tag{3.2}$$

where Hamilton defined

$$x_i = \sum_j s_{ij} b_{ij} \tag{3.3}$$

as the inclusive fitness of the ith individual. The term s_{ij} is the effect the ith individual has on the fitness of the jth recipient. The term b_{ij} is a measure of the degree to which a deviation in actor allele frequency, $q_i - \bar{q}$, predicts a deviation in recipient allele frequency, $q_j - \bar{q}$. Thus the total effect of an actor's behaviors on allele frequency change is proportional to x_i, the actor's inclusive fitness. From Eq. (3.2), we can see that an allele A that increased x_i would increase in frequency. Selection therefore increases the inclusive fitness of individuals.

HAMILTON'S RULE

Suppose an actor, i, sacrifices an amount C of its own reproductive success, and increases a recipient j's success by an amount B. Is this behavior favored by selection? Application of Eq. (3.3) allows easy calculation of the inclusive fitness effect

$$x_i = b_{ij} s_{ij} + b_{ii} s_{ii},$$

where the effect on the neighbor is $s_{ij} = B$, the effect on self is $s_{ii} = -C$, the term $b_{ii} = 1$ because the slope of an individual's genotype on its own genotype is one, and we define $b_{ij} = r$ as the coefficient of relatedness between actor and recipient. The behavior has a positive net effect when $x_i > 0$, which is

$$rB - C > 0.$$

This condition is Hamilton's rule.

The definition of relatedness requires further explanation, and Hamilton's rule requires additional justification. I will return to these topics after introducing better methods of analysis.

3.3 Queller's Quantitative Genetic Model

Hamilton's original work and the literature of the 1970s and 1980s formulated kin selection within the classical theory of population genetics. This theory typically analyzes the dynamics of a few alleles at one or two loci. Quantitative genetics provides an alternative style of analysis. The emphasis is on measurable phenotypes, such as height and weight. Measurements on populations are summarized by means, variances, and correlations among relatives. The genetics is implicit in the correlation among relatives, but one does not analyze directly specific alleles and loci.

Population genetics theory requires every detail to be specified in order to calculate the dynamics of allele frequencies. One can therefore be certain of any evolutionary deductions derived from a population genetics model, for example, whether or not an altruistic trait can increase in frequency. The price for this certitude is that one must make specific assumptions about many factors that are never known even approximately in practice. Suppose one wants to know why in some species young male birds stay with their parents to help raise siblings, whereas in other species the young males always migrate away from the nest. A population genetics analysis requires that one specify the number of loci that affect the trait, the linkage relations among the loci, the fitness of all genotypes, and so on.

Quantitative genetics has therefore dominated empirical studies that emphasize behaviors and other phenotypes. This has led to an uneasy relation between theory, often formulated in the population genetics tradition, and practical studies that depend on quantitative genetics. A few theoretical studies formulated kin selection within the framework of quantitative genetics (reviewed by Queller 1992b), and those theories generally confirmed the basic insights of Hamilton.

Queller (1992a, 1992b) demonstrated that kin selection theory has always had a close affinity to the aims and methods of quantitative genetics, in spite of the population genetics tradition for the theory. Hamilton (1970) formulated his key result with the Price Equation, which describes evolutionary change by the covariance between a phenotypic character and fitness. I discussed in Section 2.2 the relationship between the Price Equation and Robertson's secondary theorem of natural selection, which is the fundamental basis of theoretical quantitative genetics.

This close connection between the Price Equation and quantitative genetics was obscured by early studies. Both Price (1970) and Hamilton (1970) used the Price Equation, but assumed that the character of interest, z, was a gene frequency within individuals, as described in the previous section (in which the symbol q was used for z to match Hamilton's notation) . Thus Price and Hamilton could use $\overline{w}\Delta\overline{z} = \text{Cov}(w, z)$ as an exact expression for gene frequency change, requiring only Mendelian segregation so that $\Delta z = 0$. This usage transforms the Price Equation, which is generically about quantitative characters, into an exact expression of population genetics.

Queller's (1992a, 1992b) first step followed standard quantitative genetics by partitioning the character of interest into heritable and environmental components. Any character can be written as $z = g+\delta$, where g is the heritable component, or breeding value, and δ is the component not explained by genotype (Falconer 1989). The breeding value and the environmental component are uncorrelated, $\text{Cov}(g,\delta) = 0$, and the average value of the environmental effect is zero, thus $\overline{z} = \overline{g} + \overline{\delta} = \overline{g}$ (see Eq. (2.5)). This allows one to write the Price Equation as

$$\overline{w}\Delta\overline{z} = \overline{w}\Delta\overline{g} = \text{Cov}(w, g) + \text{E}(w\Delta g)$$
$$= \beta_{wg}V_g + \text{E}(w\Delta g).$$

The standard in quantitative genetics is to assume that breeding values are inherited without change between parent and offspring, so that $\Delta g = 0$. This is rarely true exactly, but provides a good approximation in most cases. Thus the standard of quantitative genetics is to live with this approximation without much concern for the consequences. This apparently innocuous assumption greatly limits the analysis of social

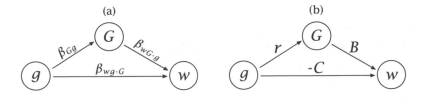

Figure 3.1   Queller's regression model for kin interactions. (a) A path diagram for Eq. (3.4). (b) The regression coefficients are relabeled with kin selection notation. The r here may differ from Hamilton's kin selection coefficient under conditions discussed later. I discuss this point in a later section. Note from the diagram that the total regression of fitness on breeding value is $\beta_{wg} = \beta_{Gg}\beta_{wG \cdot g} + \beta_{wg \cdot G} = rB - C$.

evolution. I return to this point below. Here I continue to follow Queller's development, assuming $\Delta g = 0$ and applying the shortened Price Equation, $\overline{w}\Delta\overline{g} = \mathrm{Cov}(w, g) = \beta_{wg}V_g$. The condition for the character z to increase is $\overline{w}\Delta\overline{g} > 0$ or, equivalently, $\beta_{wg} > 0$.

The next step is to partition fitness into separate causes. For social evolution, we are concerned with the direct effects of a character on fitness and the contribution of social interactions to fitness, so

$$w = \alpha_w + \beta_{wg \cdot G}g + \beta_{wG \cdot g}G + \epsilon_w, \tag{3.4}$$

where α_w is a constant, G is the average breeding value of the local group with which an individual interacts, $\beta_{wg \cdot G}$ is the partial regression of fitness on individual breeding value, holding group breeding value constant, $\beta_{wG \cdot g}$ is the partial regression of fitness on group breeding value, holding individual breeding value constant, and ϵ_w is the error term which, by least squares theory, is uncorrelated with g and G, that is, $\mathrm{Cov}(g, \epsilon_w) = \mathrm{Cov}(G, \epsilon_w) = 0$. This is a direct fitness model because w summarizes individual fitness as influenced by social partners.

Eq. (3.4) can be expanded into the path diagram in Fig. 3.1a by writing G as a regression on g, in particular

$$G = \alpha_G + \beta_{Gg}g + \epsilon_G. \tag{3.5}$$

Substituting G from Eq. (3.5) into Eq. (3.4) yields

$$w = \alpha + \left(\beta_{wg \cdot G} + \beta_{wG \cdot g}\beta_{Gg}\right)g + \epsilon$$
$$= \alpha + \beta_{wg}g + \epsilon,$$

where the expansion

$$\beta_{wg} = \beta_{wg \cdot G} + \beta_{wG \cdot g} \beta_{Gg}$$

is shown in Fig. 3.1a, and α and ϵ are combined constant and error terms from the previous regression equations. (See Li 1975, for a full discussion of regression equations and path diagrams.)

Queller matched this notation to models of kin selection (Fig. 3.1b). The direct effect of an individual's breeding value on its own fitness, $\beta_{wg \cdot G}$, determines the reproductive cost of the phenotype. To maintain the convention that cost reduces fitness, we set $\beta_{wg \cdot G} = -C$. The direct effect of average breeding value in the local group on individual fitness, $\beta_{wG \cdot g}$, measures the benefit of the phenotype on the fitness of neighbors, thus $\beta_{wG \cdot g} = B$.

The condition for the character to increase is $\beta_{wg} > 0$, which is

$$rB - C > 0$$

where $r = \beta_{Gg} = \text{Cov}(G, g)/\text{Cov}(g, g)$ is a kind of kin selection coefficient of relatedness. I discuss relatedness coefficients in later chapters.

Queller's model shows that social evolution can be studied by partitioning components of selection with multiple regression. He noted that this is an application of the commonly used multivariate analysis of selection developed by Lande and Arnold (1983), which I described in Chapter 2. With this insight, Queller realized that a broad array of regression models can be applied to social evolution, a point also made but not developed by Heisler and Damuth (1987) and Goodnight et al. (1992). Queller listed a few simple examples. Rather than repeat his particular examples, I turn to a general analysis in the next chapter. The general model clarifies many conceptual issues and provides practical techniques for the study of social evolution.

3.4 Exact–Total Models

All the models presented thus far assume that average effects are transmitted from parent to offspring without change. The models therefore have the same status as Fisher's fundamental theorem: they are exact models for partial change, holding average effects constant (see Chapter 5). In some applications it is useful to have an exact model for total change.

EXACT HAMILTON'S RULE

Starting with the exact Price Equation, Eq. (2.7),

$$\overline{w}\Delta\overline{g} = \beta_{wg}V_g + \mathrm{E}\left(w\Delta g\right),$$

the condition for the increase of an altruistic trait is $\overline{w}\Delta\overline{g} > 0$, or

$$\beta_{wg} > -\frac{\mathrm{E}\left(w\Delta g\right)}{V_g},$$

and, using the total regression of fitness on breeding value as $\beta_{wg} = rB - C$ from Fig. 3.1, we obtain an exact Hamilton's rule (Frank 1997e)

$$rB - C > -\frac{\mathrm{E}\left(w\Delta g\right)}{V_g}. \tag{3.6}$$

This is an exact, total result for all conditions, using any predictors for breeding value. The predictors of phenotype may include alleles, group characteristics, environmental variables, cultural beliefs, and so on.

Eq. (3.6) can be expressed differently by starting with the Eq. (2.8) form of the Price Equation, with g' for transmitted breeding value

$$\Delta\overline{g} = \beta_{wg'}V_{g'}/\overline{w} + D_g,$$

using the regressions $w = \alpha_w + \beta_{wg}g + \epsilon$ and $g' = \alpha_{g'} + \beta_{g'g}g + \gamma$, expanding by standard statistical definitions

$$\beta_{wg'}V_{g'} = \beta_{wg}V_g\beta_{g'g} + \mathrm{Cov}\left(\epsilon, \gamma\right),$$

and dropping the correlation of residuals, $\mathrm{Cov}(\epsilon, \gamma)$, giving

$$\Delta\overline{g} = \beta_{wg}\beta_{g'g}V_g/\overline{w} + D_g.$$

The condition for $\Delta\overline{g} > 0$ is (Frank 1997e)

$$(rB - C)\beta_{g'g}/\overline{w} > \frac{-D_g}{V_g}. \tag{3.7}$$

This form has two advantages. First, the left side shows the distinction between components of fitness, $rB-C$, and fidelity of transmission, $\beta_{g'g}$. The second advantage of this form is that it allows easy calculation, in which each term can be readily understood.

EXAMPLE: REBELLIOUS CHILD MODEL

I mentioned that the predictors used for traits can be alleles, cultural beliefs, or other variables. Here I study the evolution of a culturally inherited trait for altruistic behavior. The trait is transmitted directly from parent to offspring, but children are rebellious and switch to the opposite behavior from their parents with probability μ. For simplicity, I assume that each offspring has only one parent.

Let p be the frequency of the altruistic trait. Breeding value, g, is zero or one if the trait is, respectively, absent or present in an individual. The change in average breeding value between parent and offspring, $g' - g = \Delta g$, is μ if parental value, g, is zero, and $-\mu$ if parental value is one. The general equation for fitness, from Fig. 3.1, is

$$w = \alpha - Cg + BG + \epsilon,$$

where I have taken individual phenotype as equivalent to breeding value, g, and group phenotype as equivalent to group breeding value, G. With this setup, $p = \bar{g} = \bar{G}$, and α is chosen so that $\bar{\epsilon} = 0$.

We can obtain the equilibrium frequency of the altruistic character, p^*, when the condition in Eq. (3.7) is an equality. The terms are

$$\beta_{wg} = rB - C$$
$$\beta_{g'g} = (1 - 2\mu)$$
$$\bar{w} = \alpha + p\,(B - C)$$
$$D_g = \mu\,(1 - 2p)$$
$$V_g = p\,(1 - p).$$

This provides all the information we need to substitute into Eq. (3.7) and solve for the equilibrium frequency of altruism. The solution is a quadratic in p. When $\alpha = 0$ and $rB - C > 0$, the solution is

$$p^* = \frac{(rB - C)\,(1 - 2\mu) + \mu\,(B - C)}{(rB - C)\,(1 - 2\mu) + 2\mu\,(B - C)}. \tag{3.8}$$

Simple numerical calculations provide values of p^* for $\alpha \neq 0$. Fig. 3.2 shows how the frequency of rebellion, μ, influences the cultural evolution of altruism. Note how quickly the frequency of altruism declines when the frequency of rebellion increases from zero.

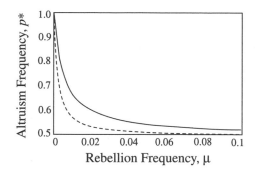

Figure 3.2   The equilibrium frequency of altruism, p^*, in a model of cultural inheritance. From Eq. (3.8) with $r = 0.1$, $B = 1.1$ and $C = 0.1$. Solid curve, $\alpha = 0$; dashed curve, $\alpha = 1$. Note that r is treated as a parameter in the spirit of comparative statics. A fully dynamic model might define r as a function of rebellion frequency, μ.

3.5 Coefficients of Relatedness

Hamilton (1970) gave a coefficient of relatedness that could be calculated from shared genealogy. His population genetic coefficient emphasized the fact that kinship causes individuals to share a common genotype by descent from ancestors. Later work, based on Price (1970), showed that statistical associations among individuals determine the course of selection (Hamilton 1972; Seger 1981; Michod 1982; Queller 1992a, 1992b). Shared genealogy is simply one process that causes statistical association.

The relatedness coefficient given above for Queller's model is

$$r = \frac{\text{Cov}\,(G,g)}{\text{Var}\,(g)} = \frac{\text{Cov}\,(G,g)}{\text{Cov}\,(g,g)}, \tag{3.9}$$

where G is the breeding value of social partners, and g is the breeding value of a focal individual. This coefficient shows the statistical nature of relatedness. Many variants on this type of coefficient have been reported, but they all have the same basic form of a regression coefficient (Michod 1982). The standard inclusive fitness coefficient takes G as the breeding value of recipients, and g as the breeding value of actors.

It is often difficult to estimate directly the statistical associations among social partners. Thus the standard approach, following Hamilton (1970), is to assume that no selection is occurring and that genes

shared from recent ancestors are the only cause of phenotypic association. Genealogy can then be used to estimate the statistical association
between partners. For example, starting with Queller's coefficient in a
diploid organism, we can write $g = x_1 + x_2$ and $G = y_1 + y_2$, where x and
y are random variables denoting the allelic effects of the two alleles at
a single locus. Assume that actor and recipient are different members
of the same population, with the social trait under study determined by
the same locus in the different individuals. Then $\text{Var}(x) = \text{Var}(y) = \sigma^2$
because x and y are alleles sampled from the same population at the
same locus. Thus

$$\begin{aligned}
\text{Cov}\,(g,g) &= \text{Cov}\,(x_1 + x_2, x_1 + x_2) \\
&= 2\sigma^2 + 2\text{Cov}\,(x_1, x_2) \\
&= 2\sigma^2\,(1 + F)\,,
\end{aligned}$$

where $F = \text{Cov}(x_1, x_2)/\sigma^2 = \text{Corr}(x_1, x_2)$ is the correlation between alleles within an individual. This correlation is often referred to as Wright's
fixation index or as the inbreeding coefficient (Crow and Kimura 1970).
A similar calculation shows that $\text{Cov}(G, g) = 4\sigma^2 f$, where $f = \text{Corr}(y, x)$
is the correlation of allelic values between different individuals. This
correlation is sometimes referred to as Wright's relatedness coefficient
(Crow and Kimura 1970).

The kin selection coefficient of relatedness can now be rewritten in
terms of the two allelic correlations

$$r = \frac{2f}{1 + F}. \tag{3.10}$$

When there is no selection, the inbreeding coefficient, F, and the allelic
correlation, f, can be calculated from a pedigree. I give examples in later
applications.

The form in Eq. (3.10) was given by Hamilton (1970) as the regression coefficient of relatedness. He used this form to connect his theory to classical population genetics. This form also highlights a "selfish
gene" interpretation of value (Hamilton 1972; Dawkins 1982). Randomly
choose one of the two alleles at a locus. That allele values its own body
by $1 + F$. The 1 is the allele's valuation of itself. The F is the probability
that the two alleles are identical. Our focal allele values another body by
the probability that partner alleles are identical to the focal allele. This
probability of identity is f for each allele in the other body, and there are

two alleles, so the valuation is $2f$. Thus r describes an allele's valuation of another body relative to its own body.

3.6 Prospects for Synthesis

This chapter began with Hamilton's (1970) classic model of inclusive fitness. Hamilton studied the population genetics of allele frequency at a single locus. He showed that an actor's net effect on allele frequency change is described in terms of its fitness effects on social partners and its statistical associations in allele frequency with those partners.

Queller (1992a, 1992b) expressed social evolution in a quantitative genetic framework. Hamilton's rule arises as a special case of the quantitative genetic model when shared genotype causes correlation of characters between social partners. Queller's study suggests a general method for multivariate analysis of correlated characters in social evolution. The next chapter develops this approach.

4 Direct and Inclusive Fitness

In spite of advances made by Queller and others, Hamilton's rule remains confusing because it is too simple to be an effective guide for many realistic problems. I clarify the concepts and methods of kin selection by placing the techniques within a broader framework for the analysis of natural selection. My extension shows that there are, in fact, two distinct processes in social evolution (Frank 1997d, 1997e).

The first process is the effect of social partners on the reproductive success of individuals. Queller (1992a, 1992b) showed that social components of selection are naturally expressed by treating the phenotypes of social partners as correlated characters of a focal individual. One can then apply the multivariate analysis of natural selection (Lande and Arnold 1983), describing interactions among kin as one process that creates correlations among characters. For example, the phenotypic correlation in sex ratio produced by two females in an isolated patch influences the favored sex ratio.

The second process is the transmission fidelity via different components of fitness. For example, a female values daughters versus nieces according to the genotypic correlation, or transmission fidelity of characters.

The difficulty is that selection and transmission are distinct processes, yet both can be described by what look to be, at first glance, the same form of Hamilton's rule. In the first case, the phenotypic correlation in sex ratio describes one common type of relatedness coefficient. In the second case, the genotypic correlation of a female to a daughter or niece describes a different type of relatedness coefficient. This creates confusion when a problem requires social aspects of selection to be separated from components of transmission. Clarification is required to understand the basic concepts of social evolution, and to use these concepts for the solution of realistic problems.

I show how correlated selection and proper weighting of transmission components can, together, provide a common currency for success. This common economic currency allows me to transform the general concepts and methods into a practical maximization technique for solving problems. (Frank 1997d, provides details of work summarized in this chapter.)

Later chapters illustrate the power of these methods applied to a wide array of problems in social evolution. The examples show that the conceptual clarifications and new techniques are essential. Many important problems cannot be solved by standard application of Hamilton's rule or Queller's extensions.

4.1 Modified Price Equation

I begin with the usual expression for a trait in terms of breeding value, $z = g + \delta$, where $\bar{z} = \bar{g}$ and $\bar{\delta} = 0$. To study the change in trait value over time, I write character value in the next time period as $z' = g' + \delta'$, with $\bar{z}' = \bar{g}'$. Thus the change in average trait value is $\bar{z}' - \bar{z} = \Delta\bar{z} = \Delta\bar{g}$. The Price Equation provides a method to obtain an exact analysis of $\Delta\bar{g}$. I use a modified version here that is convenient for my purposes.

Before proceeding, it will be useful to rearrange the scheme for indexing individuals. Social evolution is conveniently studied by dividing the population into different behavioral classes, for example, mothers, daughters, nieces, social partners, and so on. I will be concerned with individuals that are members of a particular social class. Let i index social class, and ik be members of social class i with genotype k. The frequency of the kth type in the ith class is $q_{ik} = q_i p_{ik}$, where q_i is the abundance of the ith class and p_{ik} is the abundance of the kth genotype within the ith class. The standard identities for frequencies hold, in particular, $\sum_{ik} q_{ik} = 1$ and $\sum_k p_{ik} = 1$.

I start with the simple definition

$$\Delta\bar{g} = \bar{g}' - \bar{g} = \sum q'_{ik} g'_{ik} - \sum q_{ik} g_{ik}.$$

I use the peculiar definitions of the Price Equation with respect to the indices ik. The value of q'_{ik} is not obtained from the frequency of elements with index ik in the descendant population, but from the proportion of the descendant population that is derived from the elements with index ik in the parent population. If we define the fitness of element ik as w_{ik}, the contribution to the descendant population from type ik in the parent population, then $q'_{ik} = q_{ik} w_{ik} / \bar{w}$, where $\bar{w}$ is the mean fitness of the parent population.

The assignment of breeding values g'_{ik} also uses indices of the parent population. The value of g'_{ik} is the average breeding value contributed to descendants by parents with index ik. The change in breeding value for descendants of ik is defined as $\Delta g_{ik} = g'_{ik} - g_{ik}$.

This scheme has the advantage that we can define parental classes, i, in any convenient way, and we can assign members of the descendant population to any parental class without the need to respect lineal descent. This is particularly useful for kin selection, in which one often assigns a fitness component of a neighbor to an actor whose phenotype controls the neighbor's fitness component. This will be made clear later.

We can use these definitions to write an exact expression for the change in character value

$$
\begin{aligned}
\Delta \bar{g} &= \sum_{ik} q'_{ik} g'_{ik} - \sum_{ik} q_{ik} g_{ik} \\
&= \sum q_{ik} \left(w_{ik} / \overline{w} \right) g'_{ik} - \sum q_{ik} g_{ik} \\
&= \operatorname{Cov}\left(w, g' \right) / \overline{w} + \sum q_{ik} \left(g'_{ik} - g_{ik} \right) \\
&= \beta_{wg'} V_{g'} / \overline{w} + D_g.
\end{aligned}
\tag{4.1}
$$

This matches Eq. (2.8), but the notation here is convenient for the problems of this chapter. The term D_g is the change in the effect of alleles between the parent and offspring generations. I will assume $D_g = 0$. It is important to consider what assumptions this requires

$$
\begin{aligned}
D_g &= \sum_{ik} q_{ik} \left(g'_{ik} - g_{ik} \right) \\
&= \sum_i q_i \sum_k p_{ik} \left(g'_{ik} - g_{ik} \right) \\
&= \sum_i q_i \left(\tilde{g}'_i - g_i \right) \\
&= \tilde{g}' - \overline{g}.
\end{aligned}
$$

The second line is obtained by the prior definition, $q_{ik} = q_i p_{ik}$. The third line defines the average breeding value for the character among class i parents as g_i and the average breeding value among offspring assigned to class i parents as $\tilde{g}'_i$. Note that $\tilde{g}'_i$ is defined with respect to parental frequencies, p_{ik}. Thus descendant values are weighted equally for all parents, ignoring selection and differential fitness among parents. The final line defines $\tilde{g}'$ as the average breeding value among descendants, taken with respect to parental frequencies.

Thus D_g summarizes the change in breeding value between ancestor-descendant pairs. Recall that we may assign descendants to nonlineal ancestors. Although there are many ways for D_g to be equal to zero, two general assumptions capture the main issues.

First, if there is no variation in g_i among parental classes i, then the pattern by which descendants are assigned to parental class has no effect on D_g. This assumption is reasonable when the definition of class (e.g., sister, brother) is uncorrelated with the average breeding value of the class. When breeding value for the character is associated with class definition, then the particular details of the problem should make it clear how to calculate D_g.

The second assumption to make $D_g = 0$ is that the average effect of a particular genotype does not change between parent and offspring. Changes in environment or changes in allele frequency with nonadditive allelic interactions can change the effect of genotype between parent and offspring. Changes in environment are never fully predictable. Changes in average effect can be calculated for particular assumptions about nonadditivity, but the calculations are often tedious. Average effects are approximately constant over time when the population has little genetic variance and allele frequencies change by a small amount. This constancy of additive effects is equivalent to linearization of a dynamical system within a small analytical region, the standard assumption of local equilibrium analysis.

When $D_g = 0$, the direction of evolutionary change is completely summarized by the sign of the regression coefficient, $\beta_{wg'}$ in Eq. (4.1), thus

$$\text{sign}\left(\Delta \overline{g}\right) = \text{sign}\left(\beta_{wg'}\right) = \text{sign}\left(\text{Cov}\left(w, g'\right)\right). \qquad (4.2)$$

The next section analyzes factors that influence the direction of evolutionary change.

4.2 Regression Equations

DIRECT FITNESS

This section follows Fig. 4.1 to partition the regression $\beta_{wg'}$ into components of correlated selection and kin selection. This method analyzes variation in the fitness of class i members, w_{ik}, by starting with variation in the descendant genotype of class i members, g'_{ik}. Fitness is affected by phenotypes z_{ij}, which may be controlled by other classes. The

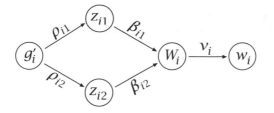

Figure 4.1   Causal chain for the association between offspring breeding value, g', and fitness, w. The pathway between g' and w is repeated for each i.

method therefore tracks the direct fitness of each class as influenced by the behavioral phenotypes of social partners.

Total fitness in the population is $W = \sum q_{ik} w_{ik}$. Each w_{ik} measures contribution to following generations for genotype k of class i, and thus implicitly includes reproductive value weightings. I discuss reproductive value in Chapter 8. Here I summarize the notation.

We can separate reproductive value and reproductive success by $w_{ik} = v_i W_{ik}$, where v_i is the reproductive value of an individual of class i. Thus total fitness is

$$
\begin{aligned}
W &= \sum_{ik} q_{ik} w_{ik} \\
&= \sum_{ik} q_{ik} v_i W_{ik} \\
&= \sum_{i} q_i v_i \sum_{k} p_{ik} W_{ik} \\
&= \sum_{i} c_i W_i.
\end{aligned}
\tag{4.3}
$$

Here the frequency of class i, q_i, is combined with the reproductive value of each member of class i as $c_i = q_i v_i$, with the v's normalized so that $\sum c_i = 1$. The c's are therefore the class reproductive values, the total contribution of class i to the following generations. The term W_{ik} is the reproductive success of genotype k of class i. The class average, W_i, is normalized to the population average W when there is no genetic variation. I assume here that each W_i refers to a fitness component with a common reproductive value weighting (see Chapter 8).

Expanding Eq. (4.2), we obtain

$$
\begin{aligned}
\text{Cov}(w, g') &= \sum_{ik} q_i p_{ik} w_{ik} (g'_{ik} - \tilde{g}') \\
&= \sum_{ik} q_i p_{ik} (v_i W_{ik}) (g'_{ik} - \tilde{g}'),
\end{aligned}
\tag{4.4}
$$

where the average effect of parental alleles in the offspring generation is $\tilde{g}' = \sum q_i p_{ik} g'_{ik}$, with the summation over parental frequencies.

Two regression equations are required to complete the paths shown in Fig. 4.1

$$W_{ik} = \alpha_w + \sum_j \beta_{ij} z_{ijk} + \epsilon_w$$

$$z_{ijk} = \alpha_z + \rho_{ij} g'_{ik} + \epsilon_z.$$

If we assume that all unspecified error terms in Fig. 4.1 are uncorrelated, then we can combine the regressions with Eq. (4.4) to obtain

$$\text{Cov}(w, g') = \sum_i q_i \left(v_i \sum_j \beta_{ij} \rho_{ij} \right) \sum_k p_{ik} g'_{ik} (g'_{ik} - \tilde{g}'). \qquad (4.5)$$

Genotypic variation can be expanded as

$$\sum_k p_{ik} g'_{ik} (g'_{ik} - \tilde{g}') = \sum_k p_{ik} g'_{ik} (g'_{ik} - g'_i + g'_i - \tilde{g}')$$

$$= \sum_k p_{ik} g'_{ik} (g'_{ik} - g'_i) + g'_i (g'_i - \tilde{g}')$$

$$= \sigma_i + g'_i (g'_i - \tilde{g}').$$

As discussed above, under the assumptions for $D_g = 0$, I assume no variation among classes, $g'_i - \tilde{g}' = 0$. When there is variation among classes, the terms $g'_i(g'_i - \tilde{g}')$ must be retained to describe selection among classes. If we drop the among-class terms and use the identity given above for reproductive value, $c_i = q_i v_i$, Eq. (4.5) becomes

$$\text{Cov}(w, g') = \sum_i c_i \sum_j \beta_{ij} \rho_{ij} \sigma_i. \qquad (4.6)$$

The term σ_i is a measure of variance among the offspring of class i, taken with respect to parental frequencies. If we assume that the distribution of genetic variance is uncorrelated with the division of fitness components into classes, then σ_i is a constant with respect to i. This is reasonable because genetic variance is often the same within behaviorally defined classes, such as sisters or brothers. If σ_i is constant with respect to i, then the direction of evolutionary change is

$$\text{sign}(\text{Cov}(w, g')) = \text{sign}\left(\sum_i c_i \sum_j \beta_{ij} \rho_{ij} \right). \qquad (4.7)$$

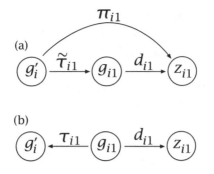

Figure 4.2   Association between offspring breeding value, g_i', and a character that influences parental fitness, z_{ij}. These diagrams expand the regression described by ρ_{ij} in Fig. 4.1. (a) Introduction of an intermediate explanatory variable, g_{ij}. The term g_{ij} is the breeding value for the character under study in an individual (actor) that influences the jth character of the ith parental class. (b) The inclusive fitness pathway. Inclusive fitness arguments usually ignore effects not associated with the actor; thus the π pathway is dropped. In order to place the actor at the center of causal explanation, the direction of the regression is changed between the actor and the recipient offspring. Thus τ is the heritability component, or relatedness coefficient, given as slope of breeding value of recipient offspring on actor breeding value.

The regression terms, β and ρ, may change with directional evolution of the character under study. Thus the condition is primarily used for describing the instantaneous direction of change for a given set of assumed or measured regression parameters or for providing equilibrium conditions.

INCLUSIVE FITNESS

Social evolution is commonly studied by inclusive fitness. The analysis begins with the individuals that control phenotype. This point of view partitions into components the ρ_{ij} regression coefficients introduced in Fig. 4.1.

Two regression equations, summarized by the diagram in Fig. 4.2a, can be combined to partition the ρ_{ij} in Fig. 4.1

$$z_{ijk} = \alpha_z + d_{ij}g_{ijk} + \pi_{ij}g_{ik}' + \epsilon_z$$
$$g_{ijk} = \alpha_g + \tilde{\tau}_{ij}g_{ik}' + \epsilon_g,$$

yielding $\rho_{ij} = \tilde{\tau}_{ij}d_{ij} + \pi_{ij}$, under the assumption that ϵ_g and ϵ_z are uncorrelated.

The ρ_{ij} of Fig. 4.1 simply summarizes the association between a phenotype, z_{ij}, and offspring genotype, g'_i, without specifying how the phenotype is determined. The g_{ij} of Fig. 4.2 is the genotype of the actor that controls the phenotype z_{ij}, which in turn influences the fitness of the recipient. The explicit role of controlling genotype introduces a natural aspect of causal analysis. The difficulty with the path in Fig. 4.2a is that the causal flow in the regression is from offspring genotype, on the left, to character, on the right, via the controlling genotype as an intermediary. Inclusive fitness takes the controlling genotype's point of view, so that both phenotype, z_{ij}, and offspring genotype, g'_i, are expressed by regressions on the controlling genotype, g_{ij}.

Fig. 4.2b shows the analysis taken fully from the controlling genotype's point of view. There are two differences from Fig. 4.2a. First, the term π_{ij} is dropped, ignoring extrinsic factors that cause an association between z and g'. Second, the direction of the regression between g_{ij} and g'_i is reversed, so that the new regression is expressed as offspring of recipient genotype on controlling genotype. This assigns variations in the abundance of descendant genotypes to the classes that control variations in phenotype. Put another way, pathways follow phenotypic cause rather than lineal descent.

When is the flip in the direction of regression valid in Fig. 4.2b? If we substitute the regressions of Fig. 4.2a into Eq. (4.6) and drop π, we have

$$\text{sign}\left(\text{Cov}\left(w,g'\right)/\overline{w}\right) = \text{sign}\left(\sum_i c_i \sum_j \beta_{ij} d_{ij} \tilde{\tau}_{ij} \sigma_i\right). \quad (4.8)$$

We use the definition of regression to switch the direction of the τ coefficient, $\tau_{ij}\psi_{ij} = \tilde{\tau}_{ij}\sigma_i$, where ψ_{ij} is the genetic variance within the class that controls phenotype j of class i. The genetic variance is always with respect to the character under study, which may differ from the phenotype z_{ij}. Substituting into Eq. (4.8) yields

$$\text{sign}\left(\text{Cov}\left(w,g'\right)/\overline{w}\right) = \text{sign}\left(\sum_i c_i \sum_j \beta_{ij} d_{ij} \tau_{ij} \psi_{ij}\right). \quad (4.9)$$

In the direct fitness formulation, I assumed that the variance among recipient offspring within class i, σ_i, is independent of i; in other words, I assumed that $\text{Var}(g'_i)$ over individuals indexed by k as constant with respect to i. Convenient analysis of inclusive fitness requires that the

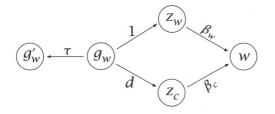

Figure 4.3 An actor has two characters, watching (w) and calling (c), that influence the fitness of a recipient. This diagram shows the inclusive fitness effect of the actor when studying the evolution of the watching character. The actor's character value for watching, z_w, has, by definition, a slope of one on its breeding value for watching, g_w. The actor's character value for calling, z_c, has a slope of d on its breeding value for watching. The value of d is influenced by linkage disequilibrium, or pleiotropy. The slope of the recipient's transmitted breeding value, g_w', on the actor's breeding value is τ, the relatedness of actor to recipient.

genetic variance within controlling classes for the character under study be the same in all classes. In particular, the variance over individuals indexed by k, $\psi_{ij} = \mathrm{Var}(g_{ij})$, must be constant with respect to ij. If one takes these variance terms as constants, then the condition in Eq. (4.9) can be simplified to

$$\mathrm{sign}\left(\mathrm{Cov}\,(w, g')\,/\,\overline{w}\right) = \mathrm{sign}\left(\sum_i c_i \sum_j \beta_{ij} d_{ij} \tau_{ij}\right). \qquad (4.10)$$

This provides the direction of evolutionary change by inclusive fitness.

Let us review the meaning of each term on the right side of Eq. (4.10). The term c_i is the class reproductive value for the ith fitness component (see Chapter 8).

The term τ_{ij} is the slope of the transmitted genotypic value g_i' on the genotypic value of the controlling class, g_{ij}. This regression is frequently defined as a kin selection coefficient (for reviews, see Michod 1982; Grafen 1985), and can also be viewed as a component measure of transmission fidelity, or heritability (Frank 1997e).

The term β_{ij} is the slope of W_i, the ith fitness component, on z_{ij}, the jth character affecting W_i. This is the commonly defined regression of fitness on multiple characters used by Lande and Arnold (1983). One difference from Lande and Arnold, however, is that here the character may be controlled by a social partner rather than by the individual itself.

The term d_{ij} is the slope of z_{ij} on controlling genotype, g_{ij}. Genotype g_{ij} is measured with respect to the character under study, which may differ from the phenotype z_{ij}. Consider, for example, guarding of social groups, in which an individual watches the periphery for predators, and gives an alarm call when a predator is sighted. The terms are illustrated in Fig. 4.3. Watching and calling are two distinct characters. When the analysis focuses on watching, then all breeding value terms, g, refer only to the watching character. The fitness of individuals is, however, also influenced by calling. One of the z's in social partners is the calling character, and the associated d term is the regression of the calling character on breeding value for watching within social partners.

COMPARISON OF DIRECT AND INCLUSIVE FITNESS

The inclusive fitness equation has advantages and disadvantages when compared with the direct fitness form in Eq. (4.7). On the positive side, the τ coefficients measure what can be thought of as components of heritability, all taken consistently from the causal actor's point of view. This is matched with the central role of actors in controlling fitnesses of the recipient classes.

On the negative side, the inclusive fitness formulation excludes π in Fig. 4.2a. This pathway measures the partial association between g' and z not mediated directly by the actor's breeding values for the character under study. The inclusive fitness formulation also requires that genetic variances be the same within all actor classes.

These problems may not be serious within the context of inclusive fitness analysis. First, if the only goal is to measure the consequences of behavior by genetic relatives, then excluding π is acceptable but may prevent complete analysis of evolutionary consequences. An example is given below. Second, many simple inclusive fitness formulations study the behavior of a single actor class; thus the requirement that genetic variances be the same within all actor classes is irrelevant.

COMPARISON WITH QUELLER'S ANALYSIS

The general models of this section differ from Queller's analysis in two ways. First, the directionality of Queller's relatedness coefficient, β_{Gg} in Fig. 3.1, is opposite to the directionality of Hamilton's inclusive fitness coefficient. Queller regresses actor genotype, G, on recipient genotype, g, providing a measure equivalent to $\tilde{\tau}$ in Fig. 4.2a. The proper measure for inclusive fitness has the direction reversed, the τ in Fig. 4.2b.

The directionality of τ is irrelevant in many common applications, as noted in the previous section. But it is important to be clear about the underlying theory and concepts.

The second difference from Queller's analysis concerns the components of transmission. Consider an actor who has a different influence on the male and female reproductive components of a recipient. In Queller's model, there is no clear way to represent these separate male and female transmission pathways. The extended model provides a distinct separation between the components of fitness and the components of transmission.

Nonadditive Models

One commonly discussed failure of Hamilton's rule arises when fitness effects are not additive. For example, suppose that fitness effects between neighbors have a multiplicative component. This can be handled in the model of Fig. 4.1 by defining a third character, $z_{i3} = z_{i1}z_{i2}$. One cannot reduce a problem to the simple $rB - C > 0$ form if benefit and cost have multiplicative effects, but the general analysis of kin selection by regression still applies (Queller 1992a).

When variation in characters is small, then multiplicative effects are approximately the sum of additive effects, $z_{i3} \approx z_{i1}\overline{z}_{i2} + \overline{z}_{i1}z_{i2}$. This separation is handled automatically by the equilibrium maximization techniques presented in the next section.

4.3 Maximization

Kin selection provides a measure of valuation, the relatedness coefficient, for comparing direct and indirect reproduction. But the theory is often difficult to apply. The regression parameters typically depend on the frequency of phenotypes and on aspects of demography. Phenotypes and demography change as selection proceeds. Thus a condition for the direction of evolutionary change, although true, is difficult to use because the parameters summarize complex processes.

The kin selection regressions above start with a complex separation of direct and indirect reproduction. It would be useful if we could turn the problem around. Begin instead with a mathematical expression that describes the direct reproductive success of individuals in terms of the biological problem of interest. Next, use standard maximization methods to obtain the direction of evolutionary change. Maximization methods provide the most powerful set of analytical tools available, and lead naturally to simple theories of comparative statics.

To use maximization, we must find a function that, when differentiated, correctly describes the direction of change under natural selection. Thus we require that differentiation automatically lead to a proper comparison of direct and indirect effects on reproduction, with an exchange rate that measures these different currencies on the same scale of value.

Marginal Direct and Inclusive Fitness

Frank (1997d) showed that a simple maximization method can be used to obtain the direction of evolutionary change (extending Frank 1995b; Taylor and Frank 1996). This allows one to start with natural, biological expressions for direct fitness given in terms of all the characters affecting each class. The solution, expressed in the form of Eq. (4.7), maintains the direct fitness point of view. The solution, in the form of Eq. (4.10), transforms the direct fitness expressions into a summary of inclusive fitness.

For direct fitness, we begin with the definition in Eq. (4.3) of total fitness, $W = \sum c_i W_i$. We then differentiate W with respect to a randomly chosen allele in the descendant population. The population is assumed to be genetically monomorphic, except for rare genotypic variants of small average effect. The method thus defines

$$\frac{dW}{dg'} = \sum c_i \frac{dW_i}{dg_i'}$$

$$= \sum_i c_i \sum_j \frac{\partial W_i}{\partial z_{ij}} \frac{dz_{ij}}{dg_i'}. \tag{4.11}$$

The left side has a natural interpretation as a slope of W on g', matching the statement in Eq. (4.2) that the direction of evolutionary change is determined by the sign of the regression $\beta_{wg'}$. The right side matches Fig. 4.1, with

$$\frac{\partial W_i}{\partial z_{ij}} = \beta_{ij}$$

$$\frac{dz_{ij}}{dg_i'} = \rho_{ij}.$$

Thus the sign of the derivative dW/dg' is sufficient for analysis of the direct fitness condition in Eq. (4.7).

A maximization method for the inclusive fitness condition in Eq. (4.10) can also be obtained. The steps for deriving the inclusive fitness form are a bit awkward, but the causal point of view of actors is often valuable. Starting with Eq. (4.11)

$$
\begin{aligned}
\frac{dW}{dg'} &= \sum_i c_i \sum_j \frac{\partial W_i}{\partial z_{ij}} \frac{dz_{ij}}{dg_i'} \\
&= \sum_i c_i \sum_j \frac{\partial W_i}{\partial z_{ij}} \frac{dz_{ij}}{dg_{ij}} \frac{dg_{ij}}{dg_i'},
\end{aligned}
\tag{4.12}
$$

where the last line matches Fig. 4.2a with

$$
\frac{dz_{ij}}{dg_{ij}} = d_{ij}
$$

$$
\frac{dg_{ij}}{dg_i'} = \tilde{\tau}_{ij}.
$$

The goal is to define a differentiation operator that leads to Eq. (4.10), without worrying about the steps that get there. Thus, following the transition from Fig. 4.2a to Fig. 4.2b, we define differentiation with respect to actor genotype by rearranging Eq. (4.12) as

$$
\frac{dw}{dg} = \sum_i c_i \sum_j \frac{\partial W_i}{\partial z_{ij}} \frac{dz_{ij}}{dg_{ij}} \tau_{ij},
\tag{4.13}
$$

where the right side matches Fig. 4.2b and Eq. (4.10). The term g_{ij} is defined as follows. Randomly choose an individual of class i, and focus on the character z_{ij} that influences the individual's fitness. Then g_{ij} is a randomly chosen actor from the class that controls z_{ij} in the focal individual. The value of g_{ij} is the breeding value of the actor for the character under study, not the genotypic value affecting the character z_{ij}.

Equilibrium is obtained by analyzing $dW/dg' = 0$ for direct fitness, or $dW/dg = 0$ for inclusive fitness, evaluated at a point with no genetic variation. Thus, at equilibrium, we take $z_{ijk} = z_{ij}^*$ in the derivative and solve for the equilibrium values of z^*, checking that the condition provides a local maximum.

Technically, the method is equivalent to the exact Price Equation, with all effects linearized by studying small variations. The term Δg is of second order (product of the change in frequency multiplied by change in phenotypic effect), and all nonlinear interactions become linear terms plus negligible second-order terms. The effect of each character on fitness is expressed as a marginal effect, $\partial W/\partial z$. Thus we obtain a marginal theory of kin selection, which is always true because fitness effects are rendered additive and selection is weak (Grafen 1985).

A locally stable equilibrium obtained by this method resists invasion by phenotypes with small deviations from the equilibrium. A locally stable phenotype is often called an Evolutionarily Stable Strategy (Maynard Smith and Price 1973; Maynard Smith 1982). Technical details of stability were discussed extensively in a special issue of the *Journal of Biomathematics* (Diekmann et al. 1996). I discuss a few aspects of dynamics in a later chapter, which uses simple games to explore the role of relatedness between social partners.

MARGINAL HAMILTON'S RULE

I provide many examples in later chapters that show the power of the marginal kin selection formulation. Here I limit myself to a derivation of the marginal form of Hamilton's rule.

Suppose that individuals are paired. One behaves as the actor, with character z. This character reduces the actor's fitness, but raises the fitness of its partner, the recipient of the behavior. This model has two classes. Class 1 is the actor, and class 2 is the recipient. Following the standard procedure, total fitness is $W = c_1 W_1 + c_2 W_2$, where the subscripts denote class, c is class reproductive value, and W_i is reproductive success. Each W_i is a function of the actor's character, z. To match Hamilton's original model, I ignore the reproductive value weightings, assuming that $c_1 = c_2$.

Marginal inclusive fitness is, from Eq. (4.13)

$$\frac{dW}{dg} = \frac{\partial W_1}{\partial z} \frac{dz}{dg_1} \tau_1 + \frac{\partial W_2}{\partial z} \frac{dz}{dg_1} \tau_2,$$

where I have dropped the j subscript because there is only one character.

The term $\partial W_1/\partial z = -C_m$ is the marginal effect of the actor's own phenotype on its fitness. To match standard Hamilton's rule notation, define C_m as the marginal cost. The term g_1 is the breeding value of the actor, who controls the character z. Following our standard model, the slope of individual phenotype on individual breeding value is one, $dz/dg = 1$. The relatedness coefficient is $\tau_1 = \mathrm{Cov}(g_1', g_1)/\mathrm{Var}(g_1)$, where g_1' is the breeding value that the actor transmits to the next generation.

The second set follows along in a similar way. The term $\partial W_2/\partial z = B_m$ is the marginal effect of the actor's phenotype on the fitness of the recipient. To match Hamilton's rule, define B_m as the marginal benefit. The term g_1 is used because the phenotype, z, is controlled by the actor class. The relatedness coefficient is $\tau_2 = \mathrm{Cov}(g_2', g_1)/\mathrm{Var}(g_1)$, where g_2' is the breeding value transmitted by the recipient to the next generation.

When variation in g is small, the condition for the character z to increase is that marginal inclusive fitness be greater than zero, $dW/dg > 0$, which yields

$$\tau_2 B_m - \tau_1 C_m > 0.$$

If we divide by τ_1, and define the kin selection relatedness coefficient as $r = \tau_2/\tau_1$, then we recover the marginal form of Hamilton's rule

$$r B_m - C_m > 0,$$

where

$$r = \frac{\mathrm{Cov}\,(g_2', g_1)}{\mathrm{Cov}\,(g_1', g_1)}.$$

This form of r is the correct definition of the relatedness coefficient for an inclusive fitness interpretation of Hamilton's rule.

This marginal version of Hamilton's rule shows that an equilibrium often satisfies the condition that marginal costs and benefits are equal, $C_m = r B_m$. Note how the relatedness coefficient, r, plays the role of currency translation between marginal effects on direct and indirect reproduction.

I show in later chapters that the marginal rule is difficult to apply directly. Instead of starting with the rule and measuring value by inclusive fitness, it is better to start with a simple expression for direct fitness of an individual. Maximization, coupled with appropriate substitution of kin selection coefficients, yields the correct answer, which is always consistent with the marginal theory of kin selection. Hamilton's rule can sometimes be used to interpret the results, but is not useful in the derivation.

4.4 Coefficients of Relatedness

Much of the literature on kin selection is concerned with justifying Hamilton's rule or describing exceptions to the rule. A large part of this literature concerns the definition of r that can make Hamilton's rule work (Michod and Hamilton 1980; Seger 1981; Michod 1982). However, many interesting problems of kin selection cannot easily be forced into an $rB - C > 0$ form. I prefer to study how the basic partitions of direct and inclusive fitness can be used to solve interesting problems.

In this section I first present a model of sex ratio evolution. This problem illustrates how direct and inclusive fitness analyses differ in their use of relatedness coefficients and phenotypic correlations. I then discuss the proper use of the transmitted breeding value term, g'. Specifically, when should this term be taken as the genotype of offspring, and when as the genotype transmitted directly by a parent through particular fitness components?

EXAMPLE: SEX RATIO

This model is the standard analysis of local mate competition, described in detail in Chapter 10. The two classes, or recipient fitness components, are male and female offspring. I use y for the sex ratio phenotype of a mother, and z for the average sex ratio phenotype of a local group. Sex ratio is the frequency of males per brood.

Each mother has fitness components for male and female offspring

$$W_m = \frac{y}{z}(1 - z)$$

$$W_f = 1 - y.$$

When there is no variation in phenotype at equilibrium, $y = z = z^*$, the fitness components have a normal value of $1 - z^*$. For direct fitness, equilibrium is analyzed by studying $dW/dg' = 0$, evaluated at $y = z = z^*$. Total fitness is $W = c_m W_m + c_f W_f$, with the c's denoting class-specific reproductive values for males and females. Differentiating the components yields

$$\frac{dW_m}{dg'_m} = \frac{\partial W_m}{\partial y}\frac{dy}{dg'_m} + \frac{\partial W_m}{\partial z}\frac{dz}{dg'_m}$$

$$= \tilde{r}_m\left(\frac{1 - z^*}{z^*}\right) - s\tilde{r}_m\left(\frac{1}{z^*}\right)$$

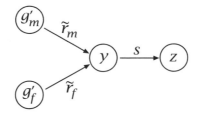

Figure 4.4   The direct fitness model for sex ratio. The analysis begins with transmitted genotypic value, g'. The associations with mothers phenotype, y, are given by $\tilde{r}$. Direct fitness also requires a measure of association between g' and average group phenotype, z. I have assumed that this association can be expressed as the product of the association between g' and y, and the association between y and z.

and

$$\frac{\mathrm{d}W_f}{\mathrm{d}g'_f} = \frac{\partial W_f}{\partial y}\frac{\mathrm{d}y}{\mathrm{d}g'_f}$$

$$= -\tilde{r}_f,$$

where Fig. 4.4 shows the new parent-offspring terms, $\tilde{r}_m = \mathrm{d}y/\mathrm{d}g'_m$ and $\tilde{r}_f = \mathrm{d}y/\mathrm{d}g'_f$, and the association between a male offspring and a random mother in the group, $s\tilde{r}_m = \mathrm{d}z/\mathrm{d}g'_m$. Solving $\mathrm{d}W/\mathrm{d}g' = 0$ yields the equilibrium for the direct fitness model

$$z^* = \frac{c_m\tilde{r}_m\,(1-s)}{c_m\tilde{r}_m + c_f\tilde{r}_f}.$$

I discuss this result after obtaining the equilibrium by the inclusive fitness method.

For inclusive fitness, the operator $\mathrm{d}g$ is interpreted as drawing a random individual from the recipient class, focusing on a phenotype that affects fitness, and picking the actor class that controls the phenotype. Analysis by inclusive fitness of $\mathrm{d}W/\mathrm{d}g = 0$ is summarized in Fig. 4.5. The analysis begins with

$$\frac{\mathrm{d}W_m}{\mathrm{d}g} = \frac{\partial W_m}{\partial y}\frac{\mathrm{d}y}{\mathrm{d}g_{my}}\tau_{my} + \frac{\partial W_m}{\partial z}\frac{\mathrm{d}z}{\mathrm{d}g_{mz}}\tau_{mz}$$

$$= r_m\left(\frac{1-z^*}{z^*}\right) - R_m\left(\frac{1}{z^*}\right).$$

The term $\mathrm{d}y/\mathrm{d}g_{my} = 1$ is the slope of individual phenotype on individual breeding value. The term $\mathrm{d}z/\mathrm{d}g_{mz}$ is the slope of a random individual on the breeding value that controls its phenotype. In this case,

(a)

(b)

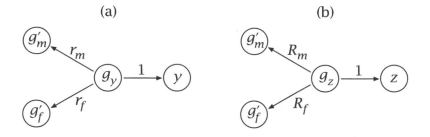

Figure 4.5   The inclusive fitness model for sex ratio. (a) The term g_y is the genotypic value of a mother, with measures of transmission (relatedness) to sons and daughters. (b) Group phenotypes are studied by choosing at random a member of the group. The term g_z is the genotype that controls the phenotype of the chosen individual. The terms R_m and R_f are the transmission measures (relatedness) of the controlling genotype to random offspring in the group.

each individual's phenotype is controlled by its own breeding value, so $dz/dg_{mz} = 1$. The term $\tau_{my} = r_m$ is the slope of transmitted genotypic value through males on maternal genotype, and $\tau_{mz} = R_m$ is the slope of a mother's transmitted genotypic value through males on a randomly chosen maternal genotype in the group. Note that R_m is equivalent to the relatedness of a mother to a random male offspring in the group.

For the female component

$$\frac{dW_f}{dg} = \frac{\partial W_f}{\partial y} \frac{dy}{dg_{fy}} \tau_{fy}$$

$$= -r_f$$

where $dy/dg_{fy} = 1$ is the slope of individual phenotype on individual breeding value, and $r_f = \tau_{fy}$ is the slope of transmitted genotypic value through females on maternal genotype. Solving $dW/dg = 0$ yields the equilibrium for the inclusive fitness model

$$z^* = \frac{c_m (r_m - R_m)}{c_m r_m + c_f r_f}.$$

The direct fitness and inclusive fitness models differ in the way genotypic value transmitted to males, g_m', is associated with aspects of group phenotype. For direct fitness, the proper measure is the slope $s\tilde{r}_m = dz/dg_m'$. This slope is group phenotype on offspring genotype. For inclusive fitness, the proper measure is $R_m = dg_m'/dg_{mz}$, the slope of offspring genotype on the genotype of a random actor in the group.

The direct fitness model is more general. For example, the phenotypes of mothers may be correlated because of common environment, behavioral coercion, shared nonadditive genetic effects, or other factors not included in the breeding value. The direct fitness expression incorporates those additional associations. In terms of Fig. 4.2a, the direct fitness model retains π, or, in terms of Fig. 4.4, the association s measures both shared breeding value and other factors.

The direct fitness analysis shows that sex ratio evolution is controlled by two factors. The phenotypic association between social partners, s, influences relative reproductive success through male and female fitness components. The genotypic coefficients, $\tilde{r}_m$ and $\tilde{r}_f$, measure transmission fidelity via male and female fitness components. Strong phenotypic associations, s, favor a low frequency of males independently of whether the association is caused by common additive genotype or by other factors.

TRANSMITTED BREEDING VALUE

I have defined g' as the transmitted breeding value—the phenotypic effect in offspring of those predictors transmitted by parents. This definition can be confusing because two alternative usages are possible depending on the context.

The first case splits each offspring into maternal and paternal components. The part of the offspring inherited from the mother is assigned to the mother. The part from the father is assigned to the father. The transmitted breeding value is the breeding value contained in gametes, when measured in the context of the offspring. Gametic breeding value is often equivalent to parental breeding value, in which case the parent-offspring relatedness coefficient is one. However, each parent is assigned only one-half of the offspring, so total valuation is one-half.

The second case assigns whole offspring to one parent only. Suppose, for example, that the genetic system is diploid and progeny are assigned to the mother. The mother's transmitted breeding value is her own gamete plus her mate's gamete. The mother's relatedness to the progeny is now the sum of two parts. She is assigned one-half of the progeny through her own gamete, and is typically related to her own gamete by one. The mother is also assigned the other half of the progeny that comes from her mate's gamete. She is related to the mate's gamete by a factor usually designated f. Her net relatedness to the progeny is $(1 + f)/2$. Thus, when whole offspring are assigned to the mother, mother-offspring relatedness is a summary statistic for several distinct phenomena.

For example, in the sex ratio model, the recipients of behavior are the mothers who produce male and female offspring. These mothers also transmit the gametic value of the their mates. Thus, a phenotype that influences the success of a mother through her sons also affects in the same way the success of the mother's mates through sons.

In the first approach, each phenotype must be evaluated for its effect on the breeding values transmitted directly by mothers, and for its effect on the breeding values transmitted directly by fathers. The second approach treats g' as the breeding value of the whole offspring, which includes the contribution from the mother and her mate. This automatically accounts for the joint effect on mothers and mates without the need to bring fathers into the analysis. The second method is commonly used in the literature, and I used it implicitly in the previous section.

The same problem arises whenever a behavior influences the fecundity of a recipient. The behavior affects both the recipient and the recipient's mates. By contrast, a behavior that influenced the mating success of a male, but not his fecundity, would have no influence on the fitness of the male's mates. In this latter case it is important to use the first definition of g' as gametic breeding value rather than breeding value of whole offspring.

5

Dynamics of Correlated Phenotypes

Many interesting biological problems concern the transition from one steady state to another. For example, most species consist of females breeding alone. A few evolutionary transitions have occurred to females breeding in cooperative groups. Comparative statics identifies the alternative steady states, but generally fails to provide a complete analysis of transitions.

My goal here is not to analyze the biology of evolutionary transitions (see Maynard Smith and Szathmáry 1995). Instead, I will describe in an abstract way the mathematics of very simple transitions. My purpose is to complement the methods of the prior chapter by studying discrete phenotypes and phenotypic variants of large effect. My main point is to emphasize, once again, the powerful role of statistical correlations in the dynamics of social evolution.

The fitness consequences of interactions are conveniently described as a game when there are two "players" (Maynard Smith 1982). The payoffs can then be displayed as a matrix. I analyze a few simple games to show how evolutionary dynamics depend on an interaction between relatedness and the range of possible phenotypes.

5.1 Games with Saddles: Peak Shifts

The matrix in Fig. 5.1 shows a simple game. Each player has two discrete behavioral options, C and D, which we can take for Cooperate and Defect. The cells show the fitness payoff to player I when it makes a particular play and its partner, player II, makes a corresponding play. For example, when both players cooperate, the payoff is a. The payoffs have been normalized in panel (b) to reduce the number of parameters to the minimal set needed to describe the game. The normalization is explained in the figure legend. I use the normalized game in (b) throughout this discussion.

Let the frequency at which player I cooperates be p, and the frequency at which its partner cooperates be q. Then the fitness of player I is

$$w_1(p,q) = pqa + (1-p)(1-q). \tag{5.1}$$

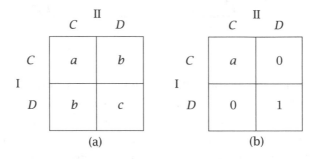

Figure 5.1   Game matrix. The cells show the payoff to player I given strategies by two players in an encounter. The game is symmetric; that is, the payoff depends only on the strategies of an individual and its partner and not on any extrinsic condition. Thus the payoff to player II can be obtained by transposing the matrix. In this case, transposition does not change the matrix. A game matrix is used to determine which strategies maximize payoffs. (a) Basic game. (b) Normalized game. The relative value of payoffs is unaffected when the same constant is subtracted from all entries, or when all entries are divided by a constant. The matrix in panel (b) is obtained from panel (a) by subtracting b from all entries, and dividing by $c - b$. The upper left entry is therefore $\hat{a} = (a - b)/(c - b)$. I assume $c > b$ and $a > b$. Although a from the left panel and $\hat{a}$ in this panel differ, I drop the hat in this panel to simplify the notation. I use the normalized game for all analyses in the text.

If partners are uncorrelated, $\mathrm{Cov}(p, q) = 0$, then the average partner frequency q is equal to the population average value of cooperation among partners, $\bar{q}$.

Fig. 5.2 shows a plot of the fitness (payoff) for player I as a function of the player's own phenotype, p, and the average partner phenotype, $\bar{q}$. The fitness surface forms a saddle, with two peaks separated by a valley. In this case, $a = 1.2$, so all individuals would do best by cooperating all the time. However, if the initial frequency of cooperation in partners is $\bar{q} = 0$, then an individual maximizes its success by always being selfish, $p = 0$. With $p > 0$, any tendency to cooperate will be strongly disfavored.

How can a population cross the valley, from the low peak at $\bar{q} = 0$, to the higher, cooperative peak at $\bar{q} = 1$? Fig. 5.3 shows that some force must move the average partner frequency, $\bar{q}$. Each point on the curve is the saddle point for a given height of the far peak, a. The value of $\bar{q}$ must be moved past the saddle point, $1/(1 + a)$, by some extrinsic force, after which selection can complete the transition.

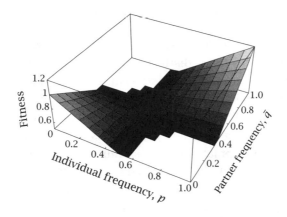

Figure 5.2   Plot of the payoffs for the game in Fig. 5.1b. The surface shows the saddle in the payoffs. A saddle point is a local maximum (stable equilibrium) in one dimension and a local minimum (unstable equilibrium) in another dimension. The saddle point occurs at $p = q = 1/(1 + a) = 1/2.4$ in this figure. This point is a local maximum along the line $p = 1 - q$ and a local minimum along the line $p = q$. This combination of stability and instability in different dimensions causes the payoff surface to have the shape of a saddle.

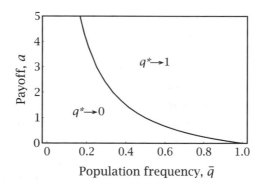

Figure 5.3   The saddle point for the game in Fig. 5.1b, for a given population frequency among partners, $\bar{q}$, and height of the far peak, a. The saddle divides the dynamics into regions in which the (C, C) peak will be attained by selection, $q^* \to 1$, and $(a, \bar{q})$ pairs for which selection moves the population to the (D, D) peak, $q^* \to 0$. It is assumed that $\bar{q} = \bar{p}$, either because both players are drawn from the same population, or because the two populations have the same average values.

5.2 Correlated Phenotypes

This peak shift problem arises in many contexts. In classical population genetics, the two players are distinct loci that interact to determine fitness. This interaction, called epistasis, is most advantageous when both loci express the C phenotype (Fig. 5.1b). Wright (1932) discussed genetic peak shifts extensively in his shifting-balance theory. Wright imagined that stochastic processes in small populations would perturb allele frequencies and allow populations to cross valleys to higher peaks. Correlations between the alleles at the paired loci can also affect the chance of a peak shift. In genetics, correlations between loci are called linkage disequilibrium.

The typical behavioral model assumes that individuals of the same population interact (Maynard Smith 1982). The relatedness of paired individuals may influence the probability of a peak shift. For example, if individuals are always paired with genetically identical partners, then playing C will increase in frequency whenever $a > 1$, regardless of the current population frequency.

The players may be different species (Frank 1994a, 1995c, 1996a, 1997c). For example, a plant and its insect pollinator may interact such that payoffs to each player depend on the joint behavior of the pair. The payoffs in a simple interaction can be described by Fig. 5.1b. Behavioral correlations between partners influence evolutionary dynamics.

Genetical, behavioral, and ecological interactions are typically treated as separate domains, requiring unique conceptual and mathematical techniques. One theme of this chapter is that statistical associations change the shape of evolutionary dynamics. The role of statistical associations are similar, perhaps identical, across a variety of traditionally distinct biological problems. We can see this by treating our basic model of fitness, Eq. (5.1), with a suitable level of abstraction. I repeat the equation here for convenience

$$w_1 (p, q) = pqa + (1 - p) (1 - q).$$

The goal is to study the evolution of p, the frequency at which player I plays the C phenotype. The frequency of the C phenotype in partners is q. The average frequency of C in the population from which player I comes is $\bar{p}$, and the average frequency in the population from which player II is drawn in $\bar{q}$. Of course, $\bar{p} = \bar{q}$ when both players come from the same population and their behaviors are controlled by the same locus.

In the two-locus genetic model, $\bar{p}$ and $\bar{q}$ are the allele frequencies at the two loci. There are three cases: (1) The two loci are expressed and interact within the same individual. This is the traditional genetic problem. (2) Each locus affects a different individual in a pair of social partners. For example, the larger partner may express p, and the smaller may express q. (3) The partners may be different species, one expressing p, and the other, q.

SMALL DEVIATIONS

Suppose individual phenotypes deviate by only a small amount from the average. We can use the simple expression for the fitness of player I in Eq. (5.1) and differentiate to determine whether fitness is increasing or decreasing with an increase in phenotype. The standard assumption for this method, outlined earlier, is that genetic variants cause only small phenotypic deviations from the population average.

Suppose the breeding value for the phenotype of player I is g_1. The derivative dw_1/dg_1', analyzed at $p = \bar{p}$ and $q = \bar{q}$, describes the rate of change in fitness for player I as its transmitted breeding value changes. For all analyses in this chapter, I assume that transmitted and parental breeding values are equal, $g' = g$. The derivative, dw_1/dg_1, is the direct fitness maximization method of Eq. (4.11), yielding the derivative analyzed at $(\bar{p}, \bar{q})$ as

$$\frac{dw_1}{dg_1} = \frac{\partial w_1}{\partial p}\frac{dp}{dg_1} + \frac{\partial w_1}{\partial q}\frac{dq}{dg_1}$$

$$N = a\bar{q} - (1 - \bar{q}) + r_1\left[a\bar{p} - (1 - \bar{p})\right],$$
(5.2)

where, as always, g_1 is normalized so that $dp/dg_1 = 1$. The term $r_1 = dq/dg_1$ is the slope of player II's phenotype on player I's genotype.

If both players are drawn from the same population, and phenotypes are determined by the same locus, then $\bar{p} = \bar{q}$. The term r_1 is the coefficient of relatedness of inclusive fitness theory only when phenotypic correlation is caused entirely by a similar genotype at the same locus.

A candidate equilibrium is obtained by solving $dw_1/dg_1 = 0$ at $\bar{p} = \bar{q} = q^*$, which yields $q^* = 1/(1 + a)$. This is the same unstable saddle point shown in Fig. 5.3, where players were assumed to be uncorrelated,

$r = 0$. Thus correlated phenotypes do not influence the dynamics. The reason is that with only small deviations from the population average, correlated phenotypes pass the saddle point only when already very near the saddle, so that dynamics are determined only by the location of the population relative to the saddle point. Put another way, prior knowledge that a partner deviates in a particular direction from its population average is of no value, because the magnitude of the deviations is too small to matter.

The two players may be drawn from different populations. If the problem is two-locus genetics, then each player is a sample from one of the two loci. If the problem is ecological, then each player is drawn from one of the two species.

Eq. (5.2) gives the change in the fitness of player I as its genotype, g_1, changes. Because the game in Fig. 5.1 is symmetric, the fitness of player II is $w_2(p, q) = w_1(q, p)$. Player II's fitness changes with the breeding value, g_2, as

$$\frac{dw_2}{dg_2} = \frac{\partial w_2}{\partial q} \frac{dq}{dg_2} + \frac{\partial w_2}{\partial p} \frac{dp}{dg_2}$$

$$= a\overline{p} - (1 - \overline{p}) + r_2 \left[a\overline{q} - (1 - \overline{q}) \right],$$

where g_2 is normalized so that $dq/dg_2 = 1$, and $r_2 = dp/dg_2$ is the slope of player I's phenotype on player II's genotype. A candidate equilibrium occurs where $dw_1/dg_1 = dw_2/dg_2 = 0$. Solving yields $\overline{p} = \overline{q} = 1/(1 + a)$, as before. This internal point is the same unstable saddle point, which does not depend on the statistical associations, r_1 and r_2. Thus, as we see in Fig. 5.2, the stable points for $(\overline{p}, \overline{q})$ are $(0, 0)$ and $(1, 1)$.

The dynamics for these assumptions are not particularly interesting. But the general problem of shifts from one equilibrium to another is important. This simple game allowed introduction of some definitions for study of dynamics, as shown in Fig. 5.4.

LARGE DEVIATIONS

What if the population is fixed at one of the peaks, with rare phenotypes from the opposite peak? When does a transition to the opposite peak occur? Correlations clearly matter. A population fixed near the lower $(0, 0)$ peak will move to the higher peak if there are rare individuals that always cooperate and always meet with cooperative partners—in other

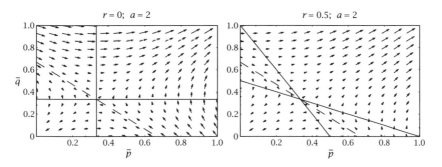

Figure 5.4   Dynamics for the game in Fig. 5.1b, with $r_1 = r_2 = r$. The arrows show the direction of change favored by selection when individuals deviate by only a small amount from their population averages, $\bar{p}$ and $\bar{q}$. Changes are obtained from the equations for dw_1/dg_1 and dw_2/dg_2 in the text. This type of graph is called a phase plot, showing the joint dynamics of variables. In the left panel, the horizontal line is the isocline for $dw_1/dg_1 = 0$. On this line, there is no change in $\bar{p}$; above the line $\bar{p}$ is increasing, and below the line $\bar{p}$ is decreasing. The vertical line is the isocline for $dw_2/dg_2 = 0$, separating regions in which $\bar{q}$ increases or decreases. The dashed line, $\bar{q} = 2/(1 + a) - \bar{p}$, separates the space into regions that attract to the $(0,0)$ equilibrium and those that attract to the $(1,1)$ equilibrium. This type of separating line is called a separatrix. In the right panel, r is increased to 0.5. The increase in r changes the shape of the dynamics, but does not change which equilibrium is attracting from a given starting point for $\bar{p}$ and $\bar{q}$.

words, when the correlation is perfect. The problem is to combine the effects of statistical associations with large deviations from the population averages.

Begin with the basic equation for fitness in this game

$$w_1\,(p,q) = apq + (1 - p)\,(1 - q)\,.$$

Next, express q as a regression that depends on deviations in p

$$q = \bar{q} + r\delta + \epsilon_q, \tag{5.3}$$

where δ can be taken as phenotypic deviation, $p - \bar{p}$ or, equivalently, as genotypic deviation, $g_1 - \bar{g}_1$. For simplicity, I assume that $p = g_1$. The term r is the regression of q on δ, and ϵ_q and δ are independent. Taking expected fitness, dropping the $\bar{\epsilon}_q = 0$ term, yields

$$w_1\,(\bar{p} + \delta, \bar{q} + r\delta) = a\,(\bar{p} + \delta)\,(\bar{q} + r\delta) + (1 - \bar{p} - \delta)\,(1 - \bar{q} - r\delta)\,.$$

Suppose that the population containing player I has two genotypes. We can write the phenotypes of these two genotypes as $p_1 = \bar{p} + \delta_1$ and

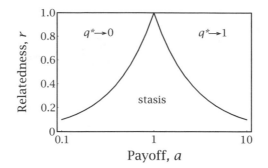

Figure 5.5   Dynamics for the game in Fig. 5.1b when the population is initially fixed for either $\bar{p} = \bar{q} = 0$ or $\bar{p} = \bar{q} = 1$, and the opposite phenotype occurs. For example, if the population is fixed at zero, and some individuals express $p = 1$ and $q = 1$, then a transition to $\bar{p} = \bar{q} = 1$ occurs only for those values of a and r shown in the upper-right region, $q^* \to 1$. Similarly, a transition to $\bar{p} = \bar{q} = 0$ occurs only for those values of a and r shown in the upper-left region, $q^* \to 0$. No changes occur in the stasis region.

$p_2 = \bar{p} + \delta_2$, with $\bar{\delta} = 0$ and $\delta_1 > 0 > \delta_2$. The condition for deterministic increase of $\bar{p}$ is

$$\Delta w_1 = E(w_1|\delta = \delta_1) - E(w_1|\delta = \delta_2) > 0,$$

which, after substitution and rearrangement, yields

$$\left(\delta_1^2 - \delta_2^2\right) r (1 + a) + (\delta_1 - \delta_2) [a\bar{q} - (1 - \bar{q}) + r (a\bar{p} - (1 - \bar{p}))] > 0.$$

When variations are small, $\delta_1^2 << \delta_1$ and $\delta_2^2 << \delta_2$, then the condition for increase matches exactly the condition for small deviations obtained in Eq. (5.2). The match occurs because differentiation is simply a technique to study $\Delta w_1/(\delta_2 - \delta_1)$ as $(\delta_2 - \delta_1) \to 0$.

When deviations are larger, we need the full expression for fitness differences with the second-order terms δ_1^2 and δ_2^2. If we start near the equilibrium $\bar{p} = \bar{q} \approx 0$, with $p_2 = 0$ and $\delta_2 = -\bar{p} \approx 0$, then the condition for increase is

$$\delta_1 > \frac{1 + r}{r (1 + a)}.$$

When the deviant type is $p_1 = 1$ and $\delta_1 = 1 - \bar{p} \approx 1$, the condition for increase is

$$r > \frac{1}{a}.$$

If both players are drawn from the same population, then $\bar{p}$ and $\bar{q}$ are equivalent, and $r > 1/a$ is the condition for transition from $(0,0)$ to $(1,1)$, as shown in Fig. 5.5. The same approach, starting at $(1,1)$, yields the condition $r > a$ for a transition to $(0,0)$. If the paired players are drawn from different populations, the dynamics of $\bar{p}$ and $\bar{q}$ must be tracked separately. The methods in the previous section can be used to study joint dynamics.

Explicit analysis of δ^2 and, in general, deviations of large effect are handled naturally with the standard regression equations for direct fitness, Eq. (4.7), or by standard application of the Price Equation. I tracked δ^2 explicitly here in order to show the connection between deviations of large effect and the calculus analysis of small deviations. (The problem of peak shifts has been discussed extensively in a variety of evolutionary and ecological contexts. See, for example, Price et al. (1993) and Mangel (1994).)

Comparative Dynamics

The results described above provide simple conditions for the increase in cooperation between partners. In this game with multiple peaks separated by a valley, mutually beneficial cooperation can be attained or prevented according to three key parameters: the height of the cooperative peak, a, the size of phenotypic deviations, and the statistical association between players.

The conclusions are simple. Transitions are easier when a is higher, phenotypic deviations are larger, initial frequencies are closer to the saddle point, and associations between players are greater. Transient statistical associations can play an important role in transitions. Once the valley is crossed, the associations are not required for maintenance of the new behaviors.

These results are comparative statements about dynamics, and might be thought of as comparative dynamics (Samuelson 1983). This approach has both costs and benefits. On the positive side, the results emphasize a few simple, biologically meaningful parameters. These parameters could potentially be measured, allowing comparative predictions to be tested. The statistical associations can be influenced by many complex processes, such as migration, recombination and selection. The simplification here shows that these diverse processes influence the probability of transition through their effect on the statistical association between partners. Thus, rather than specifying exactly how migration influences transitions, we conclude that when limited migration enhances statistical associations, it simultaneously increases the probability of transition. Put another way, I have used the statistical associations as parameters rather than dynamic variables that will change.

On the negative side, we lack the certainty of a full dynamical model. Given explicit parameters for migration, competition, population regulation, genetics, and so on, how exactly do frequencies of phenotypes change over time? Such an analysis is tedious but easy enough to do. One then has exact results in terms of many parameters. The results will not be so simple, but they can be interpreted, usually in terms of the same summary factors in the simplified model of comparative dynamics.

The problem is that the available data and the possibility of testing often do not warrant the extra detail, and one may fall into the trap of trying to fit a complex model rather than to compare alternative hypotheses (see *The Importance of Comparison*, p. 31). Also, there is no a priori reason to believe that migration, competition, and genetic control of phenotypes are themselves fixed parameters rather than dynamic variables. So one must carve between variables and parameters where the most useful insight is provided.

I continue to make the simplest cuts. The benefit is a broad and understandable overview of the logic of social evolution. The cost is that, for each problem, the theory is not fully developed into realistic models that apply to detailed cases.

5.3 Strategy Set

In the prior sections I showed how relatedness interacts with constraints on phenotype. There was, in particular, an interesting contrast between small and large phenotypic deviations. I discuss another kind of phenotypic contrast in this section—the difference between "mixed" and "pure" strategies (Maynard Smith 1982). With mixed strategies, player I has the phenotype p, which is to play the strategy C with probability p and to play the strategy D with probability $1 - p$. Player II has phenotype q, playing C with probability q and D with probability $1 - q$. With pure strategies, each individual always plays the same strategy; that is, p must be either zero or one. But the population may be a mixture of

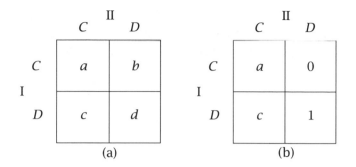

Figure 5.6   Game matrix for the general two-player, symmetric game. The cells show the payoff to player I given strategies by two players in an encounter. The game is symmetric; that is, the payoff depends only on the strategies of an individual and its partner and not on any extrinsic condition. Thus the payoff to player II can be obtained by transposing the matrix. For example, in panel (a), if player I uses C and player II uses D, then player I receives b and player II receives c. In panel (a), I assume $d > b$; that is, when an opponent is selfish (plays D), a player gets more when responding selfishly (playing D) rather than cooperatively (playing C). The game in panel (b) is normalized by starting with panel (a), subtracting b from all entries, and dividing all entries by $d - b$. In panel (b), the parameters are $\hat{a} = (a - b)/(d - b)$ and $\hat{c} = (c - b)/(d - b)$. The hats are dropped for convenience. I use the matrix in (b) in all analyses.

individuals of the two types, with an average frequency, $\bar{p}$, of strategy C.

In each example I have transcribed a game situation into an expression for fitness. The game matrix by itself is not important, except that it provides a convenient way to classify problems. I continue to use game matrices, but the approach applies generally whenever one can write an expression for fitness.

My new problem is the pairwise game in Fig. 5.6. The fitness function for player I is

$$w_1(p, q) = apq + (1 - p)(1 - q) + c(1 - p)q \qquad (5.4)$$

and the fitness function for player II is $w_2(p, q) = w_1(q, p)$.

MIXED STRATEGIES

Suppose each individual can express either strategy C or D, playing each strategy with a particular probability. As before, let player I's probability be p, and player II's probability be q. If we assume the populations are at $(\overline{p}, \overline{q})$, and only small deviations in p and q are present, then we can proceed by differentiation. The differentials are evaluated at the population averages and set to zero to obtain the values of no change

$$\frac{dw_1}{dg_1} = f(\overline{q}) + r_1 f(\overline{p}) + r_1 c = 0 \tag{5.5a}$$

$$\frac{dw_2}{dg_2} = f(\overline{p}) + r_2 f(\overline{q}) + r_2 c = 0 \tag{5.5b}$$

$$f(y) = ay - (1 - y) - yc,$$

where r_1 is the slope of player II's phenotype on player I's genotype, and r_2 is the slope of player I's phenotype on player II's genotype. These derivatives can be matched to the marginal version of Hamilton's rule. For example, the first equation is written equivalently as

$$\frac{dw_1}{dg_1} = -C_m + r_1 B_m = 0,$$

where $C_m = -f(\overline{q})$ and $B_m = f(\overline{p}) + c$.

I discuss below the case for two species or two loci, with $\overline{p}$ and $\overline{q}$ and the r's (regressions) taking on different values for the two populations. For the one-species case, in which $\overline{p} = \overline{q}$ and $r_1 = r_2 = r$, the solution is

$$q^* = \frac{r(c-1) - 1}{(1+r)(c-1-a)}, \tag{5.6}$$

where $c > (1 + r)/r$ is required for $q^* > 0$ and $c > a(1 + r)$ is required for $q^* < 1$ (Grafen 1979). From the second derivative, $c > 1 + a$ is required for the internal equilibrium to be stable, otherwise the point is an unstable saddle, and the analysis follows the discussion of the earlier section on simple games with a saddle.

The regression coefficient, r, is

$$r = \frac{\text{Cov}(P, g)}{\text{Var}(g)},$$

where g is the breeding value of an individual and P is the phenotype of its partner. This is the coefficient for kin selection models first proposed by Orlove and Wood (1978), and developed by Queller (1992a, 1992b) in his quantitative genetics formulation. The statistical association between partners does not, however, require that partners be kin or share a common genotype, although kinship is the most likely cause of association in this model.

PURE STRATEGIES

The mixed strategy assumes that a player can adopt a probabilistic phenotype, randomly expressing one strategy or another. Alternatively, the genotype may fix a player's strategy, but different genotypes may express different strategies. In the latter case, individuals are pure, but the population is mixed.

The game theory analogy is perhaps a hindrance with respect to the distinction between *mixed* and *pure*. For example, we may be interested in how an individual divides resources between sons and daughters, or dispersing and nondispersing offspring. A mixed allocation may be favored without using random expression. A pure strategy prevents an individual from splitting its allocation into two or more strategies.

A mixed-strategy equilibrium occurs when the marginal values of the different strategies are equal. A mixture of pure strategies is stable when the fitness of each hereditary particle is equal. This must be so, because if the effect of the hereditary particles is constant and the particles change in frequency, then the distribution of phenotypes changes.

The solution for the pure strategies begins with the regressions for each phenotype on breeding value

$$p = g_1 + \epsilon_p$$
$$q = g_2 + \epsilon_q,$$

where p and q always take values zero or one when the player is the focal individual. For convenience, I assume $\epsilon_p = \epsilon_q = 0$ so that I can use phenotype equivalently for breeding value.

The regressions for partner's phenotype on an individual's breeding value predict the strategy of partners

$$p = \bar{p} + r_2\delta_2 + \epsilon_2$$
$$q = \bar{q} + r_1\delta_1 + \epsilon_1,$$

where $\delta_1 = p - \bar{p}$ and has value $1 - \bar{p}$ or $-\bar{p}$ when p is zero or one. Similarly, $\delta_2 = q - \bar{q}$ has value $1 - \bar{q}$ or $-\bar{q}$ when q is zero or one.

The regressions can be used in Eq. (5.4) to obtain expected fitness under the assumption that the ϵ's are uncorrelated with each other and with breeding values. In particular, player I has expected fitness $w_1(\bar{p} + \delta_1, \bar{q} + r_1\delta_1)$, and player II has expected fitness $w_2(\bar{p} + r_2\delta_2, \bar{q} + \delta_2)$, where $w_2(p, q) = w_1(q, p)$.

Equilibrium can occur when the fitnesses for the zero and one phenotypes are the same

$$w_1 (1, \overline{q} + r_1 (1 - \overline{p})) - w_1 (0, \overline{q} - r_1 \overline{p}) = 0$$

$$w_2 (\overline{p} + r_2 (1 - \overline{q}), 1) - w_2 (\overline{p} - r_2 \overline{q}, 0) = 0.$$

If the players are drawn from the same population, then $\overline{p} = \overline{q}$ and $r_1 = r_2 = r$, yielding

$$q^* = \frac{ar - 1}{(1 - r)(c - 1 - a)}, \tag{5.7}$$

with $q^* = 0$ when $ar - 1 < 0$ and $\overline{q}^* = 1$ when $a > r + (1 - r)c$ (Maynard Smith 1982). A solution can easily be obtained when the players are from different populations, but I skip the algebra here.

This solution differs from the marginal Hamilton's rule under mixed strategies. If we release the constraint that individuals express only pure strategies, and allow individuals to express mixed strategies, then the population would usually evolve from the pure equilibrium to the mixed equilibrium.

TWO SPECIES, MIXED STRATEGIES

When partners are different species, associations cannot be caused by kinship. I develop the mixed-strategies model for two species to examine the role of statistical associations when kinship is excluded. The statistical associations play a similar role here as in the one-species model, once again showing that it is not kinship itself that determines the nature of selection.

I have described interactions in game theory language. But the fitness expression in Eq. (5.4) applies equally well to an interaction between an insect and a plant. The alternative strategies could be two different behaviors or phenotypes between which each individual divides its resources. The payoffs, or fitness function, could be made asymmetric, but I retain the symmetric model for simplicity.

The analysis is the mixed-strategy model described above, but allows $\overline{p}$ and $\overline{q}$ to differ because the traits are in separate species. The regressions, r_1 and r_2, may also differ. If we assume, for simplicity, that breeding value and phenotype are identical, $g_1 = p$ and $g_2 = q$, then

$$r_1 = \frac{\text{Cov}(q, p)}{\text{Var}(p)}$$

$$r_2 = \frac{\text{Cov}(q, p)}{\text{Var}(q)},$$

where I assume variances are small because I use the maximization method. The solution from Eqs. (5.5) is

$$p^* = (A + cr_2)/B$$
$$q^* = (A + cr_1)/B,$$

where

$$A = r_1 r_2 (1 - c) - 1$$
$$B = (c - 1 - a)(1 - r_1 r_2).$$

The solution reduces to Eq. (5.6) when $r_1 = r_2$.

The equilibrium traits for the two species differ only through r_1 and r_2. If, for example, $r_2 > r_1$, then $p^* > q^*$. From Eq. (5.4), the species with lower trait value has the higher average payoff.

Why do asymmetries in the r coefficients lead to a higher average payoff for one species than for the other? This can be understood in a generic way for symmetric fitness functions, that is, when $w_1(p, q) = w_2(q, p)$. As above, I assume that breeding values fully determine phenotype, $p = g_1$ and $q = g_2$. Marginal fitnesses are

$$\frac{dw_1}{dp} = \frac{\partial w_1}{\partial p} + r_1 \frac{\partial w_1}{\partial q}$$

$$\frac{dw_2}{dq} = \frac{\partial w_2}{\partial q} + r_2 \frac{\partial w_2}{\partial p},$$

where $r_1 = dq/dp$ and $r_2 = dp/dq$, which, for small variance, are equivalent to the statistical forms given above. If we evaluate the derivatives at $\bar{p} = \bar{q}$, then, because of symmetry,

$$\frac{\partial w_1}{\partial p} = \frac{\partial w_2}{\partial q} = -C_m$$

$$\frac{\partial w_1}{\partial q} = \frac{\partial w_2}{\partial p} = B_m.$$

Thus, at the point $\bar{p} = \bar{q}$, the marginal fitness of species 1 is $r_1 B_m - C_m$, and the marginal fitness of species 2 is $r_2 B_m - C_m$. The species with the larger regression coefficient gains a greater marginal benefit. This occurs because in the species with larger r, small increases in altruism correspond to partners with relatively larger increases in altruism. One might say that the species with the larger regression coefficient has more information, or better prediction, about its partner. This leads to an equilibrium in which the species with greater information has a higher payoff.

6 Relatedness as Information

> Correlated equilibrium is viewed as the result of Bayesian rationality; the equilibrium condition appears as a simple maximization of utility on the part of each player, given his information.
>
> —R. J. Aumann, "Correlated Equilibrium as
> an Expression of Bayesian Rationality"

Statistical associations between different loci or between different species often influence behavior in the same way as associations caused by common genealogy. Genealogical relatedness is the most important cause of associations in many problems of social evolution. But relatedness coefficients must be interpreted statistically to develop a proper theory.

The common causal chain is: genealogy → statistical association → evolutionary consequences. If a process other than genealogy causes the same statistical associations, then that process will have the same evolutionary consequences. The term *relatedness coefficient* is perhaps unfortunate because it confuses the most common cause for associations with a more general problem.

Suppose, for example, that one insists on a strictly genealogical theory. Then it would not be possible to understand why a factor, formally equivalent to the regression coefficient of relatedness, arises in ecological interactions between species. Overly narrow focus unduly limits the potential to understand the evolutionary process in a coherent way. But a purely statistical description is not satisfactory either. How can one have a theory of kin selection without a special role for genealogy?

There is no single answer because these concepts can be used for different purposes. One may, for example, wish to predict behavior for a particular genealogical structure among neighbors. Or one may wish to predict coevolution between species, given measured values for the statistical association between partners. These are different problems, but they are also problems with a close logical affinity.

I show in this chapter that information about partners is frequently a useful interpretation of statistical associations. These associations may be caused by common genealogy or by other processes.

I begin by using path analysis to clarify the meaning of statistical associations in evolutionary problems. I then turn to conditional behavior, in which each individual adjusts its phenotype in response to information about its own condition and the condition of partners. This sets the stage for the final section on kin recognition. Without kin recognition, each individual implicitly has information about the expected behavior of partners, where this expectation is summarized by the relatedness coefficient. With kin recognition, an individual actively assesses cues about the expected behavior of each particular partner. The individual then adjusts its behavior conditionally for each partner based on its assessment.

6.1 Interpretation of Relatedness Coefficients

Assume that players I and II in a game have phenotypes p and q, respectively. Then we can write the regression as in Eq. (5.3)

$$q = \bar{q} + r\,(p - \bar{p}) + \epsilon,$$
$$= \bar{q} + r\delta + \epsilon_q,$$

where $r = r_{q\delta}$ is the regression of q on p or, equivalently, the regression of q on δ, and ϵ_q and δ are independent. I define δ here as deviation of the breeding value of p,

$$p - \bar{p} = \delta + \epsilon_p.$$

Breeding value can be thought of in this context as the genetic component of phenotypic variability that is transmissible from parent to offspring (Falconer 1989). I also assume that average changes in breeding value between parent and offspring are zero, $E(w\Delta\delta) = 0$, from the Price Equation.

Fig. 6.1 describes these relations in a path diagram. The standard relatedness coefficient is the regression of q on δ, $r_{q\delta} = r$, the slope of partner phenotype on recipient genotype. This regression is shown by the line from δ to q. All of the relations in the diagram are summarized in the regression equations for p and q, and the regression equations

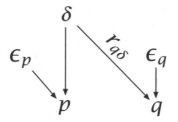

Figure 6.1   Path diagram describing the phenotypes for players I and II, p and q, and the breeding value of player I, δ. The unexplained causes, ϵ_p and ϵ_q, are independent of δ by the theory of regression analysis. I also assume that ϵ_p and ϵ_q are uncorrelated.

and assumptions about correlations among variables define the path diagram (Li 1975).

The diagram in Fig. 6.1 shows only the direct relation between δ, the breeding value of player I, and q, the phenotype of player II. It is often useful to partition this relation into parts that can be interpreted as causal components of the total association. Fig. 6.2 adds g, the breeding value of player II, so that the set of regression equations is

$$p = \bar{p} + \delta + \epsilon_p$$
$$g = r_{g\delta}\delta + \epsilon_g$$
$$q = \bar{q} + r_{q\delta \cdot g}\delta + r_{qg \cdot \delta}g + \epsilon_q,$$

where $r_{q\delta \cdot g}$ is the partial regression of q on δ when g is held constant, and $r_{qg \cdot \delta}$ is the partial regression of q on g when δ is held constant.

The regression for g can be used to expand the regression for q as

$$q = \bar{q} + \left(r_{q\delta \cdot g} + r_{qg \cdot \delta}r_{g\delta}\right)\delta$$
$$= \bar{q} + r_{q\delta}\delta,$$

where I have dropped the ϵ terms.

This algebra shows that the standard relatedness regression can be partitioned into multiple causes

$$r_{q\delta} = r_{q\delta \cdot g} + r_{qg \cdot \delta}r_{g\delta}$$

where $r_{q\delta}$ is the total regression of q on δ, from Fig. 6.1, and the right side shows the partition in Fig. 6.2. For example, in Fig. 6.2, the genotype

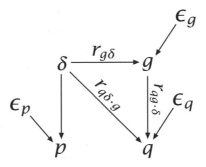

Figure 6.2 Path diagram as in Fig. 6.1, with the addition of g, the breeding value of player II. Variables not connected by a path are uncorrelated. The average values of the breeding values, δ and g, are normalized to zero.

of player I may have no direct effect on the phenotype of player II, in which case $r_{q\delta \cdot g} = 0$. Instead, both players are directly affected by their own genotypes, and there is an association between the genotypes of players I and II, $r_{g\delta} > 0$. Thus the total association of player I's genotype with player II's phenotype is $r_{q\delta} = r_{qg \cdot \delta} r_{g\delta}$. Only when $r_{g\delta}$ is caused by shared genealogy at common genetic loci is $r_{q\delta}$ the genealogical measure of relatedness frequently used in the kin selection literature.

The regression of player I's genotype on player II's phenotype, $r_{q\delta \cdot g}$, is an interesting pathway. There are several ways in which this term may be positive. For example, player I might be able to coerce player II to change its phenotype. If the genotypes of player I vary in their expressed power to coerce partners, then $r_{q\delta \cdot g} > 0$. Alternatively, player I might be able to choose a partner based on its phenotype, independently of the partner's genotype. For example, if there is a predictable environmental feature that influences player II's phenotype, then player I may exploit that information to choose partners. If player I's ability to exploit such information has a heritable component, then $r_{q\delta \cdot g} > 0$.

Regression coefficients predict the value of one variable based on the given value of a second, predictor variable. Put another way, regression coefficients describe conditional information: given the predictor, a more accurate estimate of the outcome is possible. The typical predictors are alleles, but the theory can be extended easily to use any predictors that provide information.

The models I have described show how information, measured by regression coefficients, influences evolutionary dynamics. The predictors are the genotype of an actor. The phenotypes of partners are the outcome variables that are estimated with improved accuracy.

The behaviors favored by selection are those that maximally use the information available. The information is implicitly accumulated by evolving genotypes.

Sometimes additional information is available to each individual. For example, an individual may be able to assess that it is the stronger of the two partners. I consider in the next section how behavior evolves when additional information is combined with implicit information about relatedness. In the following section I discuss how individuals may use additional information about relatedness. For example, an individual may be able to distinguish partners as full or half siblings. I analyze how direct information on relatedness can be combined with implicit information accumulated by evolving genotypes.

6.2 Conditional Behavior

Suppose each of two females starts her own nest. Label this nesting behavior N. A female may quit her nest, join the other, and contribute as a nonreproductive helper. Call this helping strategy H. The sterile worker does not reproduce and has a direct fitness of zero. Darwin noted that sterile workers posed a difficulty for his theory of natural selection. How can selection favor a trait in an individual that does not reproduce? His solution was that the workers enhanced the reproduction of individuals that carry the latent, nonexpressed tendency for sterility.

The distinction between pure and mixed strategies illustrates one type of latent expression. The game illustrated in Fig. 5.6b can be matched to the sterile worker problem. Let the Cooperate strategy, C, be equivalent to helping, and Defect, D, be equivalent to nesting alone. The payoff to player I when cooperating is always zero because cooperation is equivalent to sterility, thus $a = 0$. Player I's payoff, $c > 1$, describes its reproductive success when it reproduces with its partner helping. Nesting alone provides a payoff of one.

If strategies are pure, then an inherited tendency to be sterile causes the bearer always to be sterile. Such a trait cannot be favored, as Darwin noted. This fact is formalized by the solution for the pure-strategy equilibrium in Eq. (5.7), where the frequency of sterility is $q^* = 0$ when $a = 0$.

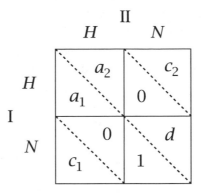

Figure 6.3   Game matrix for a two-player, asymmetric game with conditional behavior. Player I's payoff is in the lower part of each cell; player II's payoff is in the upper part of each cell. If both individuals help (play H), then there are two helpers and no reproductives. This paradox may be resolved in various ways; for example, the pair may form a helper-nester group with each individual equally likely to be in either role. This resolution yields $a_1 = c_1/2$ and $a_2 = c_2/2$.

When strategies are mixed, an inherited tendency to be sterile will sometimes be expressed, and sometimes not, according to the phenotype p. The equilibrium in Eq. (5.6) shows that, even with $a = 0$, sterility is expressed when $c > (1 + r)/r$. This can be seen by starting in a population in which no sterility is expressed. An individual that expresses a low level of sterility, δ, has partners with a level of sterility $r\delta$. This individual gains $r\delta(1 - \delta)(c - 1)$ for situations in which it nests and has a helping partner, and loses δ for the times in which it acts as a helper. The benefit is greater than the cost when $r\delta(c - 1) > \delta$, or $c > (1+r)/r$, where terms of order δ^2 are ignored because δ is small.

HELP ONLY WHEN WEAKER OF PAIR

The mixed-strategy model assumes that individuals are symmetric with respect to payoff when nesting alone, helping or receiving help. It may be that the payoffs for leaving one's own nest and helping another depend on an individual's current condition and the condition of the partner. Suppose, for example, that the conditions of the players are one and $d > 1$ for the weaker and stronger players, respectively. The payoffs for nesting alone are equal to the condition, as shown in Fig. 6.3, with player II as the stronger individual. The success of the stronger when joined by a weaker helper is c_2, and the success of the weaker when joined by a stronger helper is c_1. If both individuals help, then payoffs are a_1 and a_2 for players I and II, respectively.

Suppose, in this first model, that the stronger individual never helps. The phenotype of interest is the conditional probability, p, that an individual joins another nest as a helper, given that this individual is the weaker of the pair. If the focal individual is the stronger of the pair, then its partner helps with probability q. This assumes that individuals can assess their relative condition. As before, fitness will depend on the phenotype of our focal individual, p, and the phenotype of the partner, q. Fitness is

$$w\,(p,q) = (1/2)\,[\text{fitness if weaker of pair}]$$
$$+ (1/2)\,[\text{fitness if stronger of pair}]\,, \qquad (6.1)$$

where I assume that there is no correlation between genotype and condition, and each individual has an equal chance of being in each role. By the methods in the previous section, the direct fitness of the focal individual can be written as

$$w\,(p,q) = (1/2)\,[p\,(0) + (1-p)\,(1)] + (1/2)\,[qc_2 + (1-q)\,d]\,.$$

Suppose there are only small deviations in δ, the breeding value for p. We can find a candidate equilibrium with mixed strategies by study of $dw/d\delta$, and interpreting $dq/d\delta = r$ as the slope of partner phenotype on the genotype of the focal individual. This method shows that an individual is favored always to help when weaker, $p^* = 1$, if

$$r\,(c_2 - d) - 1 > 0.$$

Here, $c_2 - d$ is the benefit and one is the cost, so we have a simple form of Hamilton's rule. The same result holds if we allow only pure strategies.

Natural selection combines the direct information on condition with implicit information about relatedness, r. Implicit information accumulates in genotypes according to whether p is favored to increase or decrease, which is equivalent to whether r is greater than or less than $1/(c_2 - d)$.

Both Weaker and Stronger Can Help

The prior section assumed that the stronger individual never helps. When both players are potential helpers, we must track the probability that the focal individual helps when stronger or weaker, p_s and p_w, respectively, and the probability that the partner helps when stronger or weaker, q_s and q_w, respectively. Then, from the game matrix in Fig. 6.3 and the fitness expression in Eq. (6.1), the fitness of an individual is

$$w\,(p_w,p_s,q_w,q_s) =$$
$$(1/2)\,[a_1 p_w q_s + c_1\,(1 - p_w)\,q_s + (1 - p_w)\,(1 - q_s)] \qquad (6.2)$$
$$+\,(1/2)\,[a_2 p_s q_w + c_2\,(1 - p_s)\,q_w + d\,(1 - p_s)\,(1 - q_w)].$$

Here I assume that both players are drawn from the same population, so that $\bar{p}_s = \bar{q}_s$ and $\bar{p}_w = \bar{q}_w$. This reduces the problem to studying the average values of two traits, $\bar{q}_s$ and $\bar{q}_w$. It is possible to study an interaction in which each player is drawn from a separate population, requiring attention to the mean values of four traits, the probability of helping when weaker or stronger for each population.

There are two approaches for analysis. Behavioral response for each condition, in this case stronger or weaker, may be controlled by a separate trait that can potentially evolve independently of other traits. Alternatively, each individual may have a response surface; that is, behavior may be controlled by intrinsic characters that determine the quantitative response for a particular set of extrinsic conditions. I discuss these two approaches in turn.

One Trait for Each Condition

Suppose one trait controls behavior when an individual is the stronger of a pair, and a second trait controls behavior when an individual is the weaker of the pair. The problem reduces to the joint analysis of $\bar{q}_s$ and $\bar{q}_w$, the average probabilities of helping when stronger and weaker, respectively.

The statistical associations are shown in the path diagrams of Fig. 6.4. The g's are heritable predictors of the traits, with average effects that do not change between parent and offspring. The g's are normalized so that the slope of the trait on g is one. The uppercase R's are measures of

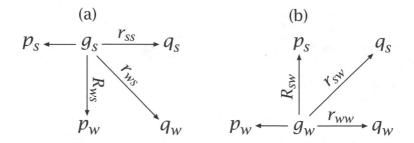

Figure 6.4   Causal pathways for two traits. (a) The stronger genotype, g_s. (b) The weaker genotype, g_w. Each variable has a single, additional ϵ cause for unexplained variation (not shown). Each ϵ is assumed to be uncorrelated with the other ϵ's and with variables besides the single variable directly affected by that ϵ. The g's are standardized so that the slope of phenotype on g is one.

association between traits within an individual. This measure is similar to the genetic covariance of traits within individuals from population genetic models. To simplify the model, I assume that $R_{ws} = R_{sw} = R$.

The associations between the same trait in different individuals are r_{ww} and r_{ss}. These are the standard kin selection coefficients, but, as always, other factors besides common genealogy may cause association. To simplify the model, I assume that $r_{ss} = r_{ww} = r$.

The association of the stronger genotype with the weaker phenotype, r_{ws}, includes direct manipulation and associations caused by correlated genetic values. I ignore manipulation and assume that these cross-associations are approximately the product of direct association and within-individual association, $r_{ws} = r_{sw} \approx rR$.

If variations in breeding values, g_s and g_w, are small, we can differentiate and replace slopes of phenotype on genotype by the appropriate coefficients. This is the direct fitness method of Eq. (4.11) with a single class. There is only one class because all genotypes have an equal chance of being stronger or weaker in an encounter. I assume transmitted and parental breeding values are equal, $g' = g$.

Differentiation is expanded by the chain rule

$$\frac{dw}{dg_s} = \frac{\partial w}{\partial p_s}\frac{dp_s}{dg_s} + \frac{\partial w}{\partial p_w}\frac{dp_w}{dg_s} + \frac{\partial w}{\partial q_s}\frac{dq_s}{dg_s} + \frac{\partial w}{\partial q_w}\frac{dq_w}{dg_s}$$

$$\frac{dw}{dg_w} = \frac{\partial w}{\partial p_w}\frac{dp_w}{dg_w} + \frac{\partial w}{\partial p_s}\frac{dp_s}{dg_w} + \frac{\partial w}{\partial q_w}\frac{dq_w}{dg_w} + \frac{\partial w}{\partial q_s}\frac{dq_s}{dg_w},$$

and substitution of the regressions for the slopes yields

$$\frac{dw}{dg_s} = \frac{\partial w}{\partial p_s} + \frac{\partial w}{\partial p_w}R + \frac{\partial w}{\partial q_s}r + \frac{\partial w}{\partial q_w}rR$$

$$\frac{dw}{dg_w} = \frac{\partial w}{\partial p_w} + \frac{\partial w}{\partial p_s}R + \frac{\partial w}{\partial q_w}r + \frac{\partial w}{\partial q_s}rR,$$

where $dw/dg > 0$, evaluated at $p_s = q_s = \overline{q}_s$ and $p_w = q_w = \overline{q}_w$, gives the condition for the increase or decrease in the associated character.

This approach takes the regression coefficients as parameters rather than dynamic variables that also change. Thus, in population genetics language, the results describe the direction of change given particular levels of relatedness and correlation of traits within individuals.

The method provides an easy way to see the shape of the dynamics as influenced by the implicit information in the coefficients, r and R, and the explicit information about condition. Specification of additional details is required to determine the complete dynamics. (Large phenotypic deviations can be studied by the methods outlined in the previous chapter.)

CONDITIONAL RESPONSE SURFACE

The previous model assumed that the response to each possible condition is encoded by a separate character. That is plausible if the number of alternate conditions is small. In the prior model the conditions were stronger or weaker, so two characters were sufficient.

Often it makes sense to describe phenotype as a functional response to condition. For example, the probability of helping, p, may be written as a function of condition, x,

$$p(x) = \sum_{i=0}^{n-1} \alpha_i x^i,$$

where there is a set of n characters, $\{\alpha_i\}$, that determines the functional response to condition. In a pairwise interaction, the probability that a partner with condition y helps is

$$q(y) = \sum \beta_i y^i.$$

If the fitnesses of player I and II are $w_1[p(x), q(y)]$ and $w_2[p(x), q(y)]$, then the dynamics of small deviations can be studied from dw_1/dg_i for

$i = 0, \ldots, n - 1$, where g_i is breeding value for α_i, and from dw_2/dh_i for $i = 0, \ldots, n - 1$, where h_i is breeding value for β_i. This will lead to consideration of the statistical associations among the $\{\alpha_i\}$ and $\{\beta_i\}$.

6.3 Kin Recognition

The previous models described associations by regression coefficients. Those models were sometimes vague about which individuals interacted. Instead, the coefficients were the average similarities over the set of individuals that interact in a particular context.

Suppose, for example, that siblings interact within a nest. Let a particular pair of parents have diploid genotypes at a locus, AB and CD, where each of the alleles is rare in the population. Each offspring is equally likely to be one of AC, AD, BC, or BD. An individual offspring, with two alleles at this locus, has full siblings with an average number of one matching allele. The average match frequency is one-half, which is also the average genetic relatedness in this case. But one-quarter of siblings will have two matches, with perfect identity to the focal individual, and one-quarter will have no matches.

If the focal individual cannot discriminate the number of matches, then behavior evolves according to the average relatedness. The average relatedness, given the context of the nest, is one-half. The context, in the absence of further discrimination, provides a prior distribution of genetic similarities. The standard models of kin selection assume average relatedness based on the context-dependent distribution.

What if an individual can discriminate the number of matches, and therefore the exact similarity at this particular locus? The problem of kin recognition is a natural extension of relatedness as information. The problem can be restated as follows. The context of the interaction provides prior information on the expected similarity between partners. Expected similarity is often one-half within a nest, and zero when there is random mixing of partners in the population. Does an exact match at a particular locus change the expected similarity between partners? If so, an individual can adjust its behavior conditionally based on a direct estimate of relatedness.

The matching locus presumably codes for a chemical or visual cue that can be detected. But this locus itself does not directly influence the behavior under study, such as the tendency to be altruistic. How

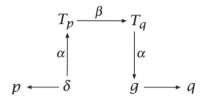

Figure 6.5   Information about statistical associations of behavioral characters provided by matching indicator traits. The only link between the indicators and behaviors is the association (linkage) within individuals, α. The variables p and q are the phenotypes of the focal individual and partner, respectively, with breeding values δ and g. Each individual has associated indicator traits, T_p and T_q. The breeding values and codings for the indicator trait are standardized to have a mean of zero and variance of one, and the slope of phenotype on breeding value is one. The statistical association between the focal individual's genotype, δ, and the partner's phenotype, q, is $\alpha^2 \beta$, where α^2 is the association within individuals between behavioral and indicator traits, and β is the association between indicator traits of partners. A small association, α, between the behavior and the indicator trait, provides essentially no information because the overall association depends on α^2.

much information does a sensory match provide about similarity of the behavioral traits?

The problem can be split into two ways in which the match provides information about behavioral similarity (Crozier 1987; Grafen 1990). First, the match provides information when there is a statistical association between the indicator characters and the behavioral characters. Second, the match provides information when there is a common, extrinsic process that determines similarity between partners in matching characters and in behavioral characters. For example, common genealogy may cause similarity at all loci.

INDICATOR AND BEHAVIORAL TRAITS

Suppose individuals express indicator traits, such as distinctive odors or colors. If an individual and its partner match for rare indicator traits, how much information does this match provide about statistical associations of behavioral traits? The information depends on the degree to which matching indicators imply matching behavioral traits. In particular, the information about a behavioral match depends on the square of the association between the indicator and behavior (Fig. 6.5).

The association between indicator traits and behavioral traits is probably low in most cases. Statistical associations between developmentally unrelated traits occur only when there is an association, or linkage disequilibrium, between genes at different loci. Disequilibrium is low because recombination breaks up associations among loci. Association is preserved only when some extrinsic force directly maintains the disequilibrium.

Matching of behavioral traits occurs by a sequence of associations in Fig. 6.5. An alternative mechanism is commonly discussed in the kin recognition literature. Suppose an indicator trait, such as a green beard, is associated with the behavioral tendency to be cooperative (Dawkins 1982). Then, in the sort of games discussed earlier, individuals are favored to recognize cooperative partners by their green beards, and reciprocate the cooperative behavior. This is an unstable situation. Uncooperative individuals would often gain by expressing a green beard, taking advantage of mistaken recognition by partners. This would destroy the association between green beards and cooperative behavior (but see Haig 1996).

COMMON GENEALOGY

Matches at indicator loci provide information about associations of behavioral traits when a common process creates association at all loci (Grafen 1985). The standard cause of genetic similarity is common genealogy. All parts of the genome that are inherited in the same way are drawn from the same distribution of shared genealogical ties. Between a pair of individuals that share recent ancestors, the average association of traits is the same for all traits. Indeed, the correlation of traits is the standard way to estimate the genealogical relationship of individuals, under the assumption that the traits are not affected by selection (Fig. 6.6).

Consider a haploid dominant life cycle in order to keep the problem simple. Cells from haploid adults act as gametes. These fuse to produce a diploid zygote, which undergoes standard meiotic processes to produce haploid adults. Thus two siblings, produced from the parental zygote, inherit the same copy of an allele with probability one-half.

Let there be n indicator loci. Each locus has $1/p$ different alleles of frequency p. For a pair of individuals, a match occurs at a locus if both individuals have the same allele. The total number of matches across

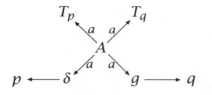

Figure 6.6   Information about statistical associations of behavioral characters provided by matching indicator traits. Common descent from an ancestor, A, provides the only link between the indicators and behaviors. Otherwise, the diagram is the same as Fig. 6.5. The association between indicator traits, T_p and T_q, is a^2, which is equal to the association between the behavioral traits, p and q. Data that provide an estimate of the association between indicator traits also provide an estimate of the association between behavioral traits.

the n loci is m. The problem is to estimate, for a pair of individuals, the statistical association between alleles at a behavioral locus, given that there are m matches at the indicator loci. Assume no association between indicator loci or between an indicator locus and the behavioral locus.

The probability of a match at any locus is

$$q = f + (1 - f)\, p, \tag{6.3}$$

where f is the correlation coefficient between alleles at the locus, which, in the context of this haploid model, is the coefficient of relatedness. It is also useful to note that

$$f = \frac{q - p}{1 - p}. \tag{6.4}$$

The correlation is caused by shared ancestry. The probability of m matches follows a binomial distribution with parameters n and q

$$P(m) = \binom{n}{m} q^m (1 - q)^{n-m}. \tag{6.5}$$

If m matches are observed, the best (maximum likelihood) estimate for q is $\hat{q} = m/n$. Since p is a fixed parameter, the best estimate for f, from Eq. (6.4), is

$$\hat{f} = \frac{\hat{q} - p}{1 - p}. \tag{6.6}$$

The variance of $\hat{f}$ is, from the standard statistical theory for binomial estimators

$$\mathrm{Var}\left(\hat{f}\right) = \mathrm{Var}\left(\frac{\hat{q} - p}{1 - p}\right) = \frac{q\,(1 - q)}{n\,(1 - p)^2}, \tag{6.7}$$

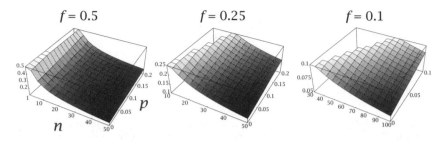

Figure 6.7 Information about common genealogy based on matches at indicator loci. The height in each graph is twice the standard deviation of the estimator, $\hat{f}$, calculated from Eq. (6.7). The variation depends on the number of indicator loci, n, the allele frequency at these loci, p, and the true value of correlation caused by common genealogy, f. Each panel is arranged to emphasize conditions under which twice the standard deviation of the estimator is less than the true value.

where q can be expressed in terms of f and p from Eq. (6.3). The graphs in Fig. 6.7 show the amount of information about f that can be obtained from n indicator loci.

CONTEXT AND INDICATORS: BAYESIAN ANALYSIS

Consider behavioral interactions among chicks in a bird's nest. A chick might interact with full siblings, half siblings that have a different father, and unrelated chicks. Nonrelatives may be placed in the nest by unrelated mothers or by cuckoo (parasitic) species. If a chick cannot discriminate different degrees of relatedness to particular nestmates, then its behavior will evolve according to the average relatedness in nests. Put another way, the context of sharing a nest provides prior information about expected relatedness.

An adult bird, after it disperses away from its parents, will encounter mostly nonrelatives. Perhaps rarely it will encounter, by chance, a cousin or more distant relative. The context of distance from birthplace provides prior information about the probability of meeting relatives of different degree.

Typical kin selection models use expected relatedness based on context. The assumption is that individuals cannot discriminate particular relatives based on kin recognition. In contrast, a typical kin recognition model assumes no prior information from context. The assumption is that each individual conditionally adjusts its behavior for each partner based on estimated relatedness from matching loci. The model of kin recognition in the previous section is of this type.

It seems plausible that an animal would combine prior information from context with additional recognition information from matching loci. For example, Getz (1982) studied discrimination between full and half siblings. He implicitly assumed that the probability is one-half for encountering each type of sibling. Often, from context, there is a greater degree of prior information about the probability of encountering each type of relative.

How can prior information be combined with matching data for each partner? The standard analytical approach is to use the rules of conditional probability. When these rules are used to combine prior information with new data, the method is called Bayesian inference (e.g., Lindgren 1976).

The key rule of conditional probability can be expressed by considering two events, X and Y, as

$$P(XY) = P(X|Y)P(Y),$$

which is read as: the joint probability that two events occur, X and Y, is equal to the probability that X occurs given that Y has occurred, multiplied by the probability that Y has occurred. Using this same rule, we can also write

$$P(XY) = P(X|Y)P(Y) = P(Y|X)P(X)$$

which can be rearranged into Bayes's theorem

$$P(Y|X) = \frac{P(X|Y)P(Y)}{P(X)}.$$

This is a general statement about how to combine information. Suppose, for example, that Y is the event that a partner is a brother, and X is a particular number of matches at indicator loci. The prior probability from context that Y is a brother is $P(Y)$. The probability of observing X matches, given the assumption that Y is a brother, is $P(X|Y)$. The posterior probability that Y is a brother, given that one has observed X matches, is $P(Y|X)$. The denominator on the right side is simply the sum over all possibilities in the numerator, so that the right side is a proper probability

$$P(X) = \sum_Y P(X|Y)P(Y).$$

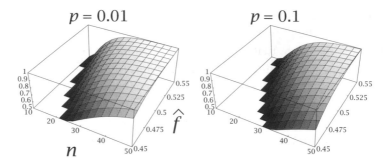

$p = 0.01$ $p = 0.1$

Figure 6.8   Discrimination between full and half siblings by combining prior expectations from context with observed matches at indicator loci. The height of each plot shows the posterior probability that a partner is a full sibling, calculated from Eq. (6.8). The prior probabilities are $a = 0.1$ for full sibs, $f = 0.5$, and $a = 0.9$ for half sibs, $f = 0.25$. The observed matches, m, are expressed in terms of the estimated relatedness, $\hat{f}$, under the assumption of no prior information (see Eq. (6.6)).

The interpretation of Bayes's theorem for kin recognition can be restated as follows, using $m = X$ for number of matches, and $f = Y$ for genetic correlation between partners (relatedness). The new estimate for the probability that a partner is a brother, $f = 1/2$, given m matches at the indicator loci, is the prior probability from context, $P(f)$, multiplied by the probability of observing m matches under the assumption that the partner is a brother, $P(m|f = 1/2)$. This product is normalized by $P(m)$ over all prior probabilities for relatedness obtained from context.

I illustrate the combination of prior and current information with the following setup. Let the prior distribution for relatedness be

$$P(f = f_1) = a$$
$$P(f = f_2) = 1 - a.$$

The probability of m matches, given a value for f, is, from Eq. (6.5),

$$P(m|f) = \binom{n}{m} q^m (1-q)^{n-m},$$

where q was defined in Eq. (6.3) as

$$q = f + (1-f)\,p,$$

$p = 0.1$

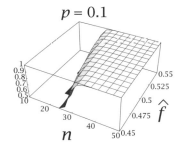

Figure 6.9 Bayesian discrimination between full sibs and nonrelatives. The height of the plot shows the posterior probability that a partner is a full sibling, calculated from Eq. (6.8). The prior probabilities are $a = 10^{-6}$ for full sibs, $f = 0.5$, and $1 - a$ for nonrelatives, $f = 0$. See Fig. 6.8 for comparison and details.

and the binomial parameters, n, for number of loci, and p, for allele frequency at each locus, are set extrinsically. Thus, from Bayes's theorem

$$P\,(f = f_1|m) = \frac{P\,(m|f = f_1)\,P\,(f = f_1)}{\sum_f P\,(m|f)\,P\,(f)}. \tag{6.8}$$

Discrimination between full and half siblings is often discussed with respect to social insect colonies, in which a worker might discriminate between siblings that have the same or different fathers. For example, in honey bee colonies, the single queen typically mates many times. Most interactions among workers are between half siblings, but some pairs are full siblings (Breed and Bennett 1987).

Discrimination between full and half siblings in a haploid organism is illustrated in Fig. 6.8. The data from matches are often sufficient to provide an estimate of relatedness, $\hat{f}$, that is close to one-half, and significantly different from zero, according to Fig. 6.7. Yet, given the prior probability of 0.1 that a partner is a full sibling, the posterior probability from the matching data is often below 0.5, indicating an inference that the partner is likely to be a half sibling. This shows the powerful role played by context in the discrimination of kin.

The problem of discriminating full and half siblings in a family is often more difficult than picking a full sibling from a crowd of unrelated individuals. Fig. 6.9 shows that picking one sibling among one million nonrelatives is easier than the nestmate recognition problem in Fig. 6.8.

Another problem, emphasized by Grafen (1985), is the discrimination of distantly related individuals from a crowd of mostly nonrelated individuals. Suppose, for example, that the probability of encountering a partner related by 0.05 is about one in a million in a randomly mixing population. Calculations along the lines above show that it is, indeed, very difficult to recognize a relative with $f = 0.05$ when among unrelated individuals, even when many matching loci are used.

Polymorphism at Matching Loci

Recognition by matching alleles is improved when alleles are rare, that is, when p is small. Recognition systems, where they occur, probably use polymorphisms that are maintained for other reasons, such as the vertebrate MHC alleles involved in immunity.

Can polymorphisms be maintained if they are only used in kin recognition systems? Grafen (1990) suggested that polymorphism for matching is maintained by selection on the recognition system. The argument is easiest to describe when matching loci have only one common allele and one rare allele. Suppose that individuals use information about the number of matches independently of how many common and rare alleles they possess. Those individuals with more rare alleles than average will obtain more accurate information about relatedness from the number of matches with a partner. If information improves fitness, then rare alleles will increase in frequency. Continual selection favoring rare alleles will lead to the maintenance of many rare alleles at each locus.

Does more accurate information about relatedness of partners improve fitness? Suppose an individual interacts equally frequently with partners related by r_1 and partners related by r_2, with average relatedness $\bar{r} = (r_1 + r_2)/2$. Let fitness be determined by the generic two-player game in Fig. 5.6. Assume, initially, that there is no discrimination, so that all players use the prior estimate of $\bar{r}$. Then, if mixed strategies are allowed, the equilibrium behavior evolves to the frequency of cooperation given in Eq. (5.6).

Consider a rare individual that discriminates relatives of class r_1 from those of class r_2. Can this individual increase its fitness by conditional behavior when compared with the population that tends not to discriminate and behaves uniformly according to the average relatedness, $\bar{r}$? Grafen (1990) has argued that discrimination does allow this individual to increase its fitness. If it did not, then kin recognition would be of no value. It seems likely that Grafen's argument is correct, simply because conditional behavior based on additional information is likely to increase fitness. But no one has yet written down a full model to analyze this problem, and the associated consequences for polymorphism of matching alleles.

6.4 Correlated Strategy and Information

The payoffs in social interactions change when partners have correlated behaviors. The measure needed turns out to be the regression coefficient of partner on individual. This regression measures information about the behavior of partners or the transmissible genotype of partners. Shared genealogy is a common cause of association between partners, but other factors can cause association.

Information is accumulated by the evolution of genotypes based on a repeated context for social behavior. Additional information about likely payoffs is available when an individual can assess factors such as resources or the attributes of specific partners. This conditional information may be combined with the implicit information from context. Bayesian inference is a natural method for combining prior and context-dependent information.

The notions of correlated strategy and information in social evolution have a close affinity to game theory models with correlated behaviors (Aumann 1974, 1987; Skyrms 1996). The goals of the game theory literature differ from those presented here. That literature is particularly concerned with the philosophical interpretation of rationality and inference. But the relationship to my analysis can be seen by repeating the quote from Aumann's classic paper with which I began this chapter: "Correlated equilibrium is viewed as the result of Bayesian rationality; the equilibrium condition appears as a simple maximization of utility on the part of each player, given his information" (1987, 1).

7 Demography and Kin Selection

The marginal costs and benefits of social acts depend on the demographic context. For example, if total reproduction in a group is strictly limited to a fixed number, then increasing one's own reproduction necessarily reduces the reproduction of neighbors. By contrast, dispersing away from the local group reduces local pressure on resources and enhances the reproduction of neighbors.

I describe the role of demography with four examples. The first two address the problem of strictly limited local reproduction. In the third example, the evolution of a character influences the availability of resources. The fourth model has separate fitness components, which are affected differently by changing demography. This prepares the way for the following chapter on reproductive value, the formal theory for assigning weights to different fitness components.

7.1 Viscous Populations

The spatial scale of population regulation influences the evolution of altruism in a general way. Consider, for example, the expression for fitness

$$w\,(y,z,\bar{z}) = \frac{bz - cy}{az\,(b - c) + (1 - a)\bar{z}\,(b - c)}, \tag{7.1}$$

in which an individual invests y in altruistic acts, at cost cy to itself. The average level of altruism in the neighborhood is z, with beneficial effect bz on fitness. The focal individual's reproduction is therefore proportional to $bz - cy$, which is the numerator.

The denominator is the intensity of competition for scarce resources, which increases as the average reproductive success rises. The average reproduction in the neighborhood is the local average of $bz - cy$, which is $z(b - c)$ because the local average of y is z. The average in the population is $\bar{z}(b - c)$.

The parameter a is the spatial scale of density-dependent competition. An increase in the reproductive success of neighbors by a proportion δ increases local competition by a factor $a\delta$. An increase in the average reproductive success of the population by a proportion y increases global competition by a factor $(1-a)y$.

Eq. (7.1) can be analyzed by the usual direct fitness method, dw/dg', where g' is a small deviation in transmitted breeding value. If we replace the phenotypic derivatives by the appropriate coefficients, the condition for the altruistic character to increase is

$$\frac{dw}{dg'} = \frac{\partial w}{\partial z}\frac{dz}{dg'} + \frac{\partial w}{\partial y}\frac{dy}{dg'}$$
$$= rB_m - C_m$$
$$= r[b - a(b-c)] - c > 0,$$

where $B_m = \partial w/\partial z$ and $C_m = -\partial w/\partial y$ are the marginal benefit and cost for an increase in altruistic behavior. The term $r = dz/dg'$ is a type of kin selection coefficient. The usual assumption, $g' = g$, yields the standard coefficient $r = \text{Cov}(z,g)/V_g$.

When population regulation is completely local, $a = 1$, then altruism cannot spread (Wilson et al. 1992; Taylor 1992a, 1992b; Queller 1994). Altruistic behavior cannot enhance the reproduction of neighbors in this case because neighborhood reproduction is strictly limited by resources. When population regulation is global, $a = 0$, then the condition for increase is $rb - c > 0$. The distinction between local and global regulation emphasizes the need to consider marginal costs and benefits in the full context of behavior and demography (Goodnight 1992; Kelly 1992). The direct effects on reproduction, b and c, are often different from the marginal effects on net reproduction.

Sex ratios provide a particularly interesting example of local and global regulation. In some cases, males compete locally against other males for access to mates (local mate competition), whereas females disperse and compete globally for resources. This tends to diminish the marginal gain for allocation to males relative to females, and thus to bias the sex ratio toward females.

In other cases, males disperse and compete in the mixed population, whereas females compete locally for resources (local resource competition). This tends to diminish the marginal gain for allocation to females relative to males, and thus to bias the sex ratio toward males. I discuss these examples in Chapter 10.

7.2 Dispersal in a Stable Habitat

Local regulation increases competition among relatives and reduces the potential benefits of cooperation. This tension may be partly resolved if individuals disperse away from their relatives and compete with non-relatives. The key parameter is the scaling of dispersal compared with competition, in particular, the average dispersal distance compared with the average distance between individuals that compete for a limited resource.

This interaction between competition and dispersal is now well understood in terms of kin selection. The development of these concepts played an important role in the study of demography and kin selection. I summarize the history in four stages.

HAMILTON AND MAY'S MODEL

Hamilton and May (1977) asked: why do many organisms disperse from their site of birth even though the probability of death during dispersal is high? Their setup is simple. Assume a habitat has a large number of discrete sites that can support a particular species. In each year, the parents die after producing babies. Each baby has a trait that determines the probability, d, that it disperses from its natal patch. Those that stay at home, with probability $1 - d$, compete for one of N available breeding sites. Dispersers die with probability c, and with probability $1 - c$ they find a patch in which to compete for one of the local breeding sites. All sites are occupied in the simple model discussed here. Hamilton and May (1977) analyzed the case in which one ($N = 1$) breeding site is available in each patch.

Hamilton and May (1977) assumed, in their first model, that the organism is asexual. They found the Evolutionarily Stable Strategy (ESS) dispersal rate, d, to be

$$d^* = \frac{1}{1 + c},\qquad(7.2)$$

where c is the cost of dispersal. Their paper contained few analytical details. A reconstruction and extension of their method is given below.

Their second model analyzed a sexual organism. This raises difficulties because in the patch structure of the model, the organism is likely to be inbred to some extent. Inbreeding was not easily handled by Hamilton and May's (1977) methods. If the males disperse freely before mating, so that the population is outbred, then the ESS dispersal rate for females is

$$d^* = \frac{1 - 2c}{1 - 2c^2} \qquad 0 < c < 1/2$$
$$= 0 \qquad 1/2 < c < 1. \tag{7.3}$$

MENDELIAN ANALYSIS

Motro (1982a, 1982b, 1983) used a fully recursive genetic model to study the same problem. He wrote equations for the fitness of each genotype as a function of the biological assumptions outlined above and the frequencies of other genotypes. Motro also specified whether the dispersal trait of offspring is controlled by the mother or by the offspring itself. This model has fully explicit assumptions about genetics and dynamics. ESS models, such as Hamilton and May's, require fewer assumptions but are sometimes more difficult to interpret.

Motro found that, in an asexual model, his result agreed with Hamilton and May, $d^* = 1/(1 + c)$. Motro obtained two additional results. First, in a sexual model in which the mother controls the dispersal trait of offspring, the same $d^* = 1/(1 + c)$ result occurs. By contrast, when Motro tried to match the assumptions of Hamilton and May for offspring control of phenotype, he obtained

$$d^* = \frac{1 - 4c}{1 - 4c^2} \qquad 0 < c < 1/4$$
$$= 0 \qquad 1/4 < c < 1, \tag{7.4}$$

which differs from Hamilton and May's result in Eq. (7.3). Motro drew two conclusions. First, in sexual models, the equilibrium depends on whether offspring phenotype is controlled by the mother or by the offspring. Second, under offspring control, explicit population genetic models failed to confirm Hamilton and May's result, which Motro attributed to a failure of the simplified ESS method.

ANALYSIS BY KIN SELECTION

Motro's different results for parent versus offspring control suggested parent-offspring conflict over dispersal. Such differences between parent and offspring typically arise because a parent is equally related to each offspring, whereas an offspring is more closely related to itself than to its siblings. Thus a parent will tend to treat all offspring equally, whereas an offspring will be inclined to favor itself over its siblings (Trivers 1974).

The hint of parent-offspring conflict in this dispersal model suggested a role for kin selection. But the prior models did not use kin selection as an analytical tool. At that time it was not clear how to incorporate kin selection into such models; instead, kin selection was often invoked after the fact to explain the outcome of an analysis.

I developed a Price Equation method of analysis (Frank 1986a). Rather than using kin selection as an explanation for results of complex models, this method allowed kin selection to be used as an analytical tool. The method reduces complex breeding systems and demographies to a simple problem of maximization. Here I summarize Frank (1986a) but use the direct fitness maximization method (see Section 4.3), which provides a simpler approach. The first step is to write the fitness of an individual in terms of its dispersal probability d, the average phenotype of individuals in its patch, d_p, and the average phenotype in the population, $\bar{d}$, as

$$w\left(d, d_p, \bar{d}\right) = (1-d)\, p\,(d_p) + d\,(1-c)\, p\left(\bar{d}\right),$$

where

$$p\,(\alpha) = \frac{1}{N\left(1 - \alpha + \bar{d}\,(1-c)\right)}$$

is the probability that an offspring competing on a z patch will win a breeding spot. Each parent produces a large number of progeny.

Using the direct fitness maximization method, we obtain a solution by assuming the population is at an equilibrium and studying the fitness effect of variants in transmitted breeding value, g'. If variations in breeding value are small, then the equilibrium is found by solving $dw/dg' = 0$ and defining r as the derivative dd_p/dg'. The candidate equilibrium occurs when the marginal cost for investing a little more in dispersal equals the marginal benefit for the reduced competition at home, that is, when

$$\frac{dw}{dg'} = -C_m + rB_m = 0.$$

where, for this dispersal model

$$C_m = \frac{c}{1 - cd}$$

$$B_m = \frac{1 - d}{(1 - cd)^2},$$

yielding the ESS

$$d^* = \frac{r - c}{r - c^2} \qquad 0 < c < r$$

$$= 0 \qquad r < c < 1. \qquad (7.5)$$

Here the equilibrium condition is consistent with the marginal version of Hamilton's rule. But the solution itself was obtained by writing a biological expression for direct fitness, and then using r as an exchange rate for value to combine direct and indirect fitness effects on the same scale. It would have been difficult to start with the marginal version of Hamilton's rule and derive the equilibrium result.

This solution for dispersal polymorphism subsumes the prior results of Motro and of Hamilton and May. All of their results are for only one breeding site per patch, $N = 1$. If the organism is asexual, then the coefficient of relatedness among individuals in the patch is one, and we recover the result in Eq. (7.2). If the organism is sexual, and offspring phenotype is controlled by the mother, then $r = 1$, and we again have Eq. (7.2). The reason $r = 1$ with parental control is that we are considering parental phenotype, and when $N = 1$, the parent competes only with itself, to which it is related by one. This can be seen formally by noting that, from the methods outlined in Section 3.5,

$$r = \frac{\text{Cov}(d_p, g)}{\text{Cov}(d, g)} = \frac{\text{Cov}(d_p, g)}{V_g},$$

where I have assumed that $g' = g$. Here d is individual phenotype and d_p is group phenotype. If the phenotype of interest is the mother's and there is $N = 1$ mother in each patch, then $d = d_p$.

If, by contrast, offspring control their own phenotypes and are outbred, then relatedness among competitors is $r = 1/2$ because competitors are siblings, and we recover Eq. (7.3). In this case, d is the phenotype of an offspring, and d_p is the average phenotype of that offspring's full siblings. Motro implicitly assumed that the mother mated several times

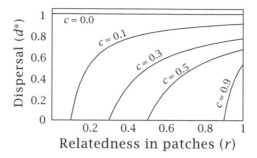

Figure 7.1 ESS dispersal when the coefficient of relatedness among competitors on a patch, r, is used as a parameter.

and offspring in a patch were only half sibs, so in his model $r = 1/4$, leading to Eq. (7.4). Thus the single kin selection model explains the parent-offspring conflict and the difference between Motro's analysis and Hamilton and May's model.

With the kin selection model, we are not limited to one breeding site, $N = 1$, or to an outbreeding system. Rather, we can treat r as a parameter and express the ESS dispersal fraction in terms of the coefficient of relatedness (Fig. 7.1). Higher relatedness increases dispersal. The reason is that an allele competing with close relatives gains little by winning locally against its relatives (Frank 1986a). Even a small chance of successful migration and competition against nonrelatives can be favored.

DEMOGRAPHIC ANALYSIS

The solution $d^* = (r - c)/(r - c^2)$ presents an interesting puzzle. Dispersal depends on the coefficient of relatedness, r, and r depends on dispersal probability, d. This coupling occurs because the relatedness among individuals in a patch depends on the frequency of successful migration. The assumption here is that shared genealogy is the only cause of correlation among social partners.

Taylor (1988a) showed that r can be expressed in terms of d and other demographic parameters. This allows the solution to be "unwound" and expressed so that d depends only on independent demographic parameters. Taylor's methods are important because the coupling between relatedness and demography is common in models of social behavior.

Taylor began with standard genealogical calculations from population genetics to obtain allelic correlations within groups. These correlations

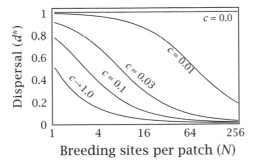

Figure 7.2   ESS dispersal for a haploid, asexual model, given explicitly in terms of the demographic parameter N.

depend on the genetics, mating system and migration scheme. The simplest case is haploid, asexual genetics, such that the allelic correlation among breeding females on the patch is given by the recursion

$$F' = (1/N) + F\,[(N-1)/N]\,(1-m)^2$$

where the correlation at a particular time, F', has two components. The first is the probability, $1/N$, that two alleles sampled randomly from the group come from the same breeding female. The second component is the probability, $(N-1)/N$, that the alleles come from different individuals who, with probability $(1-m)^2$, were both born in that patch, and that those different alleles sampled from the same patch were correlated in the previous time period by F. The value of m is the effective migration rate, so $1-m$ is the probability that an adult was born in the patch in which it breeds. The effective migration rate is

$$m = \frac{(1-c)\,d^*}{1-cd^*}. \tag{7.6}$$

The equilibrium for the allelic correlation is obtained by setting $F' = F$, yielding

$$F = \frac{1}{N - (N-1)\,(1-m)^2}.$$

In a haploid model, the kin selection coefficient is, at equilibrium, equal to the allelic correlation between individuals, $r = F$. Taylor (1988a) showed that one can use this equilibrium value of r in the solution for the ESS dispersal rate in Eq. (7.5). Since r depends on dispersal, d^*, the

ESS condition from the substitution of r into Eq. (7.5) is a polynomial in d^* that depends on the cost of dispersal, c, and the number of breeding adults per site, N. Thus the coefficient of relatedness, r, is eliminated, and the solution for d^* can be expressed entirely in terms of the demographic parameters c and N (Fig. 7.2).

The calculation and interpretation of genealogical coefficients is simple in haploid, asexual models. The structure of the problem is the same for complex genetic systems and patterns of competition, dispersal, and mating, but the calculations are more tedious (Frank 1986b, 1986c, 1987c; Taylor 1988b).

SUMMARY OF DISPERSAL ANALYSIS

The solution for dispersal in Eq. (7.5) in terms of the coefficient of relatedness, r, applies to many kinds of genetic systems. For example, one may assume haploidy or diploidy and various mating patterns. Other forces besides shared genealogy may also cause correlations among social partners. The generality is gained by leaving the genetic details unspecified and expressing the association among competitors on each site by a generalized relatedness coefficient. This coefficient is a statistical association that, in turn, depends on the unspecified genetic and demographic details.

If one wishes to obtain a solution for specific demographic and genetic assumptions, then the equilibrium kin selection coefficient can be obtained by standard techniques of genetic recursion. This allows the statistical associations of kin selection to be expressed in terms of genealogy. The genealogical calculations provide an accurate description of the statistical kin selection coefficients only when selection is very weak. This weak selection condition is satisfied by analyzing the population at its genetically monomorphic equilibrium and by considering the selection of mutations that deviate from this equilibrium by small amounts.

The solution of dispersal in terms of relatedness is also a useful result in itself, rather than solely a point of departure for further demographic analyses. Suppose, for example, that we have estimates of relatedness in different populations. The solution can be used to predict the direction of change in the dispersal rate as a function of relatedness. Here the nature of the data allows relatedness to be used as a parameter that predicts changes in dispersal.

Primacy of Comparative Statics

The dispersal result is given as an exact numerical prediction for equilibrium. The equilibrium depends on parameters affecting relatedness and the cost of dispersal. Such numerical predictions for equilibrium are common in the theoretical literature. This often leads to testing and inference following from the mistaken notion that the goal is to obtain a fit between theory and observation: if one measures the parameters, does the observed dispersal rate match the predicted dispersal rate? If the match is not good enough, by whatever criteria, one must ask whether parameters as complex as cost of dispersal or benefit of social interaction were measured with sufficient accuracy. Usually little confidence can be placed on the definition or measurement of such complex terms.

The proper use of the theory is comparative. Among populations with different r, the theory predicts that dispersal will increase with relatedness. In symbols, $\partial d^*/\partial r > 0$, which expresses the formal notion of comparative statics by analysis of how an equilibrium is expected to change as a parameter changes.

The kin selection coefficient, r, summarizes many demographic processes. For any demographic parameter, α, the influence of that parameter on dispersal can be understood via its effect on relatedness. Formally, if the parameter does not influence the cost of dispersal, $\partial c/\partial \alpha = 0$, then the sign of $\partial d^*/\partial \alpha$ equals the sign of $\partial r/\partial \alpha$.

Comparative predictions for change in equilibrium value form an approach called *comparative statics*. Comparative expressions do not require that all populations be at equilibrium. Rather, if the simplified theory properly distills crucial processes of the system, then the comparative prediction describes how the system tends to change as a single parameter changes, holding other parameters constant. Comparative predictions have the advantage that one need only measure the direction of change in parameters and characters.

7.3 Joint Analysis of Demography and Selection

The statistical association between actor phenotype and recipient genotype captures in a fundamental way the causal processes of selection. When correlated phenotypes interact, any formulation that does not summarize succinctly those statistical associations will turn out to be analytically cumbersome and difficult to interpret.

The dispersal problem is relatively easy because the demographic condition, the number of breeding females in each patch, is a fixed parameter. However, demographic properties, such as density, often vary in response to the character under study. A joint analysis of selection and demography is required.

I illustrate these issues with a model of cytoplasmic incompatibility caused by bacterial infection (Frank 1997a). Cytoplasmic incompatibility influences the fertility of mating between different kinds of individuals. The demographic character of the population is the frequency of host individuals infected by the bacteria. The bacterial character under selection is the reduction in host fertility for particular kinds of pairings.

CYTOPLASMIC INCOMPATIBILITY

Wolbachia are maternally inherited infections found in many insects. These bacteria sometimes cause incompatibility between infected and uninfected mates. A cross between an infected male and an uninfected female is sterile, whereas all other crosses are fertile. This form of sterility is commonly called cytoplasmic incompatibility (for reviews, see Rousset and Raymond 1991; Werren et al. 1995; Clancy and Hoffmann 1996).

To analyze the evolution of incompatibility, the first step is to write a fitness function that describes how biological assumptions influence reproduction

$$w(y, z) = \frac{(1 - a - by)(1 - \mu)}{(1 - q)^2 + q(1 - a - bz) + q(1 - q)(1 - z)}, \tag{7.7}$$

where the fitness of a parasite in a female, w, depends on the parasite's trait value, y, and the average value of this trait among neighbors with which the infected female interacts, z.

The parasite trait under study, when in a female host, reduces fecundity by an amount by. All infected females have their reproductive rate reduced by a; thus the focal female's reproductive rate is proportional to $1 - a - by$, as in the numerator. The parasite is vertically transmitted, and there is only one parasite genotype in each host. The probability of transmission is $1 - \mu$; that is, an infected mother has a fraction μ of her offspring uninfected. Thus the reproductive rate of a parasite is equal to the reproductive rate of its host female multiplied by $1 - \mu$. Transmission probability, $1 - \mu$, is uncorrelated with the level of incompatibility, y.

I assume that population regulation occurs within neighborhoods. Our focal female's fecundity must therefore be compared with the average fecundity in the neighborhood, given in the denominator of Eq. (7.7). The frequency of infected individuals is q, and I assume that this frequency is the same in both males and females.

Given those assumptions about frequency of infection, the frequency of matings between an uninfected male and an uninfected female is $(1 - q)^2$, and the relative fecundity of the uninfected female is one. Mating pairs with an infected female occur with frequency q, and the relative fecundity of infected females in the neighborhood is $1 - a - bz$. Matings between infected males and uninfected females occur at frequency $q(1 - q)$. The trait under study causes incompatibility in these matings. The average value of the trait in the neighborhood is z, so the average fecundity of uninfected females mating with infected males in the neighborhood is $1 - z$.

DEMOGRAPHY INDEPENDENT CASE

If $b = 0$, incompatibility has no correlated effect on female fecundity. The direction of change in incompatibility favored by selection can be determined by the sign of dw/dg' evaluated at the population average $y = z = \bar{z}$. I assume throughout that the bacterial population has only small variants in phenotype about $\bar{z}$. Fitness, w, is given in Eq. (7.7), and g' is transmitted breeding value. For this model, I assume transmitted and parental breeding values are equal, $g' = g$. Differentiation is straightforward, and the condition for $dw/dg > 0$ is

$$rq(1 - q) > 0, \tag{7.8}$$

which shows that selection favors an increase in incompatibility whenever $r > 0$ and there is some polymorphism in infection status ($q \neq 0, q \neq 1$). Here r is the relatedness coefficient, $r = \mathrm{Cov}(z, g)/V_g$.

The result in Eq. (7.8) shows that kin selection favors an increase in the level of incompatibility when population regulation occurs at the neighborhood level. The reason can be seen by inspection of Eq. (7.7). An increase in the incompatibility trait, y, by an amount δ, is associated with an increase in the average incompatibility of neighbors, z, by an amount $r\delta$. If $b = 0$, a rise in z increases fitness whenever there is any polymorphism in infection status.

The frequency of infection, q, depends on the level of incompatibility (see below). However, Eq. (7.8) is sufficient to show the direction of evolutionary change in incompatibility whenever there is polymorphism in infection status.

FIXED DEMOGRAPHY

Incompatibility causes a correlated decrease in the fecundity of infected females when $b > 0$. This case is interesting because the beneficial effects of incompatibility, proportional to the coefficient of relatedness in the neighborhood, r, must be compared to the direct reduction in female fecundity, b. The condition for selection to favor an increase in the average value of the incompatibility trait, $\bar{z}$, is obtained by the same methods as above, yielding

$$rq\,(b + 1 - q) > b \tag{7.9}$$

when the values of a and b are small relative to one, as expected in real situations.

The frequency of infection, q, is the demographic context for selection. I have treated q as a fixed parameter in this case. This makes sense for the following examples of comparative statics. Given measured values of q that differ among populations, what is the comparative prediction for incompatibility among populations? What change in incompatibility would be predicted within a population if some extrinsic force changed demography?

VARIABLE DEMOGRAPHY

Suppose one wishes to make comparative predictions as the pleiotropic effect of incompatibility, b, changes. Then one must account for the joint effects of the parameter b on both incompatibility and demography.

To put the matter formally, changes in incompatibility, $\bar{z}$, will influence the infection frequency, q. Simultaneously, q influences the natural selection of $\bar{z}$. The effect of q on $\bar{z}$ is expressed in Eq. (7.7). The effect

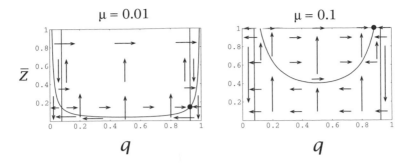

Figure 7.3   Joint dynamics of the frequency of infected individuals, q, and the average level of incompatibility, $\bar{z}$. These plots were made from Eq. (7.11) and Eq. (7.12). The parameters used are $a = 0$, $b = 0.005$, and $r = 0.07$. The two vertical lines separate regions in which $\bar{z}$ is favored to increase or decrease. The curve separates regions in which q is increasing or decreasing.

of $\bar{z}$ on q can be expressed as the condition for an increase in q

$$(1 - a - b\bar{z})\,(1 - \mu) > (1 - q)^2 + q\,(1 - a - b\bar{z}) + q\,(1 - q)\,(1 - \bar{z}).$$
$$(7.10)$$

The left side is the number of infected progeny produced by an infected female, and the right side is the average number of progeny produced by all females. The left and right sides are, respectively, the numerator and denominator from Eq. (7.7), ignoring genetic variation in the incompatibility trait so that $y = z = \bar{z}$. I ignore genetic variation because I assume throughout that the population is genetically monomorphic except for rare variants of small effect. I also assume that the frequency of infection, q, is the same in all subpopulations.

Let $a = 0$ to highlight the relative roles of b and r. Assuming that b and μ are small relative to one, Eq. (7.9) and Eq. (7.10) can be written as

$$q^2 - q\,(1 + b) + b/r < 0 \qquad (7.11)$$
$$q^2 - q\,(1 + b) + b + \mu/\bar{z} < 0, \qquad (7.12)$$

where the top inequality sets the condition for an increase in $\bar{z}$ and the bottom inequality sets the condition for an increase in q. These inequalities allow one to sketch a phase plane for the joint dynamics of $\bar{z}$ and q (Fig. 7.3).

The equilibrium $z^* = q^* = 0$ is locally stable. An internal, locally attracting point may also exist when $r > 4b$. There are two cases. If

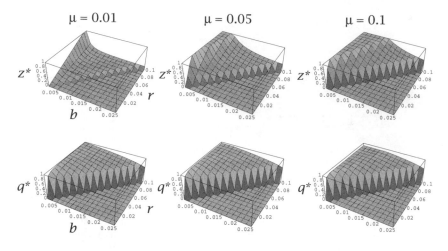

Figure 7.4   Internal equilibrium values of the frequency of infected individuals, q^* and the level of incompatibility, z^*.

$r < b/(b + \mu)$, then there is an internal equilibrium at

$$z^* = \frac{r\mu}{b\,(1 - r)}$$

$$q^* \approx \frac{1 + \sqrt{1 - 4b/r}}{2}.$$

This situation is shown in the left panel of Fig. 7.3. If $r > b/(b+\mu)$, then

$$z^* = 1$$
$$q^* \approx 1 - b - \mu.$$

This type of equilibrium is shown in the right panel of Fig. 7.3. Comparison of the left and right panels of Fig. 7.3 shows that equilibrium incompatibility, z^*, increases as the transmission efficiency of the bacteria, $1 - \mu$, decline. As the bacteria are lost more frequently from hosts, higher μ, the polymorphism in infection status, $q^*(1 - q^*)$, increases. Incompatibility is advantageous only in matings between infected males and uninfected females; thus an increase in the polymorphism of infection enhances the benefit of high incompatibility.

The internal equilibrium, when it exists, is shown in Fig. 7.4. The infection tends to be absent or at high frequency, as shown for the equilibrium value, q^*. The level of incompatibility varies over a wide range, influenced by all three parameters, b, r, and μ. The transmission efficiency, μ, appears to cause the most pronounced effects on the level of incompatibility. This analysis is intended only as a rough, qualitative guide to the complex dynamics of this system. The main point is that relatedness, r, can strongly influence the selection of incompatibility.

COMPARATIVE PREDICTIONS: VARIABLES AND PARAMETERS

Often one is interested in how given changes in demography influence the direction of change in the selected character. In that case, demography can be treated as a fixed parameter, and the analysis is relatively simple. In other cases, the comparative predictions of interest concern a parameter, such as the pleiotropic effect of incompatibility, that influences jointly the selected character and demography. Then one must treat demography as a variable and conduct a joint analysis. The method of small variations often allows simple treatment of selection (Charlesworth 1994).

Large variations in phenotype are important when the question concerns peak shifts—the movement from one stable equilibrium to another. In this case, the regression method for *Large Deviations*, p. 84, is often sufficient. The goal would be to predict changes in the likelihood of peak shifts as some underlying parameter changes.

7.4 Components of Fitness

Many social problems are conveniently described by separating two distinct components of fitness. The first is the relative success of an individual compared with its social partners. The second is the average success of the group of social partners. An individual's total success is the product of these two components. I describe this separation formally. I then apply the theory to a model of individual versus group success and to a model of parasite virulence. The latter model shows how demography influences the relative weighting of fitness components. The proper weighting, called reproductive value, is the topic of the following chapter.

Suppose fitness can be written as

$$w(y,z) = I(y,z) G(z),$$

where y is the phenotype of the focal individual, z is the phenotype of that individual's social partners, I is individual fitness relative to the average fitness of social partners, and G is the average fitness of social partners.

I study this model with the standard approach of setting transmitted breeding value to parental breeding value, $g' = g$, and then examining small variations in g. The value of g is normalized so that $dy/dg = 1$. Thus $r = dz/dg$ is the slope of average partner phenotype on individual breeding value, and is a common form of the kin selection coefficient. Evaluating dw/dg at $y = z = z^*$ yields

$$\frac{dw}{dg} = I_y G + r (I_z G + I G_z) \qquad (7.13a)$$

$$= -C_m + r B_m, \qquad (7.13b)$$

where the subscripts denote partial differentiation with respect to that variable. From the marginal version of Hamilton's rule, the result shows that marginal costs are $C_m = -I_y G$ and marginal benefits are $B_m = I_z G + I G_z$.

An individual is part of its social group in this formulation. If there are n individuals in a group, and the individuals are mutually unrelated, then $r = dz/dg = 1/n$. This association occurs because an individual is perfectly correlated with itself and uncorrelated with the other $n - 1$ members. In general, the total value of r can be partitioned into $r = 1/n + r'[(n-1)/n]$, where r' is the average pairwise relatedness between the focal individual and the other $n - 1$ members of the group.

TRAGEDY OF THE COMMONS

Suppose there is a common pool of resources available to a social group. Each individual increases its own reproduction relative to neighbors by taking more resources. However, as each individual takes more resources for itself, the average productivity of the group declines.

The model

$$w(y, z) = \frac{y}{z}(1 - z)$$

is perhaps the simplest example of this tension between individual and group success (Frank 1994b). In this model $I = y/z$, showing that an individual's relative success depends on its relative phenotypic value.

This term can be thought of as competitiveness compared with neighbors. The group term, $G = 1 - z$, shows that average success declines as average competitiveness in the group rises. Low values of z correspond with prudent use of resources and high sustainable yield.

The general solution in Eq. (7.13a) can be applied to obtain a candidate equilibrium, yielding

$$\frac{dw}{dg} = \frac{1 - z^*}{z^*} - r\left(\frac{1}{z^*}\right) = 0,$$

and the solution is $z^* = 1 - r$.

Equilibrium fitness is $1 - z^*$. Both individual and group do best when trait values are low. Low competitiveness and prudent use of resources are not stable, however, unless there is a strong coupling between individual and group success, measured by r.

The instability of cooperation illustrates the famous "tragedy of the commons," apparently first described in formal economic terms by William Forster Lloyd (1833; see Hardin 1993). The tragedy is that each individual gains by pursuing interests that increase returns relative to neighbors and decrease the value of common goods. Shared resources tend to be overexploited, to the detriment of both individual and group. The tragedy can be overcome only when individual and group success are tightly coupled.

Parasite Virulence

A parasite's fitness depends on the rate at which it transmits progeny to new hosts and on how long its current host survives. The transmission component is influenced by the availability of susceptible hosts. Host availability is, in this case, a demographic variable, because it describes the distribution and abundance of the parasite.

I analyze this model in three steps. First, I describe the dynamics of host availability. Second, from this demographic model, I write an expression for parasite fitness in terms of individual and group components. Fitness depends on a character that enhances individual transmission rate but decreases host (group) survival. Third, I analyze this model with standard kin selection tools to find the joint evolutionary and demographic equilibrium.

A common demographic model for host availability is

$$dU/dt = \theta - \delta U - \beta(y)\,US$$
$$dS/dt = S\,(\beta(y)\,U - \delta - z - c)$$

where U and S are, respectively, the number of uninfected hosts available and the number of sick hosts that cannot be infected again (Anderson and May 1991). The total population is maintained by new, uninfected hosts that are recruited at a rate θ. For each contact between uninfected and sick hosts, an individual parasite transmits progeny at a rate $\beta(y)$, where y is the transmission character of the individual parasite. Parasites die in a sick host when the host dies or the infection is cleared. The clearance rate is c. The natural host death rate is δ. The parasites increase this death rate by the virulence level, z, where z is the average value of y within the host. When variation in character values is small, so that $y \approx z \approx \overline{z}$, then the demographic equilibrium occurs at

$$U^* = (\delta + \overline{z} + c)\,/\beta(\overline{z}).$$

The fitness of an individual parasite can be extracted from this demographic model as

$$w(y, z) = I(y, z)\,G(z),$$

where

$$I(y, z) = \beta(y)\,U$$
$$G(z) = 1/(\delta + z + c),$$

where I is the fecundity per time unit of an individual parasite, and G is the expected survival time of that parasite. There is a "tragedy of the commons" tradeoff between fecundity and survival (Frank 1996b). Higher values of y enhance transmission from one host to another because $\beta(y)$ is an increasing function of y. Virulence, z, is the average value within a host of the individual parasite's transmission character. Thus the greater the fecundity of each individual, the more quickly the host resources are depleted. Host death causes parasite death.

The equilibrium condition for the parasite character is obtained directly from Eq. (7.13a)

$$\frac{dw}{dg} = I_y G + rIG_z = 0$$

evaluated at $y = z = z^*$, with $I_z = 0$. Candidate equilibria occur at the solution for z^* in

$$\frac{\beta'(z^*)}{\delta + z^* + c} - r\left(\frac{\beta(z^*)}{(\delta + z^* + c)^2}\right) = 0, \qquad (7.14)$$

where $\beta'(z^*)$ is the partial derivative of $\beta(y)$ with respect to y evaluated at z^*. Particular solutions and applications of this model are discussed in Frank (1996b) and further developed in Section 8.3.

The equilibrium does not depend on the demographic variable, U, the number of uninfected hosts available. This is surprising at first glance. It would seem that the more hosts available, the greater weight would be placed on transmission over survival. This would increase transmission and virulence. However, in this model the joint demographic and evolutionary equilibrium occurs at the point that equalizes the rate of return on the fecundity and survival components of fitness (Bull 1994; Lenski and May 1994; Frank 1996b).

If there were an oversupply of uninfected hosts, with the infection spreading in epidemic fashion, then the fecundity component of fitness would be given extra weight (May and Anderson 1990). This accounts for the value of capturing a larger share of the new resource by enhancing transmission at the expense of survival. If the supply of infected hosts were below equilibrium, then the survival component of fitness would be given extra weight. In this case, it pays to maintain current resources, the current host, rather than invest in the exploitation of new resources by transmission to rare, uninfected hosts.

The proper weights for different fitness components is called reproductive value. The next chapter shows how to use reproductive value in models of social evolution.

8

Reproductive Value

> [I]f we regard the birth of a child as the loaning to him of a
> life, and the birth of his offspring as a subsequent repayment
> of the debt ... at what rate of interest are the repayments the
> just equivalent of the loan?
> —R. A. Fisher, *The Genetical Theory of Natural Selection*

A behavior often affects more than one individual. We have seen that the coefficient of relatedness provides the proper exchange rate with which to measure multiple consequences in a single currency. This standardized measure of valuation allows one to use simple tools of maximization.

Coefficients of relatedness are sufficient when all individuals value themselves equally. But an old, sick individual may have essentially no chance of future reproduction, whereas its child may be in the prime of life. Similarly, a newborn has many dangers to pass before it can expect to reproduce, whereas its mother is usually in her reproductive prime. How should one weight these differences in expected reproduction?

The common currency in evolutionary studies must always be the magnitude of change caused in the future composition of the population. Thus one should weight each individual by its expected contribution to the future of the population. Fisher (1958a) defined this weighting as reproductive value.

Kin selection and reproductive value provide the currency translations needed to measure behavioral consequences on a single scale. This has been recognized for many years. But, just as with the simple kin selection models, the full power of maximization techniques has not been used to study the combined effects of kin selection and reproductive value. This chapter summarizes the basic tools.

8.1 Social Interactions between Classes

The simplest use of reproductive value provides an extended Hamilton's rule

$$rBv_p - Cv_a > 0, \qquad (8.1)$$

where B and C have their usual interpretations as marginal changes in reproductive success. The reproductive value of the social partner is v_p; the reproductive value of the actor is v_a.

In one sense, it is obvious that this extended Hamilton's rule is true. The valuation of a benefit to a partner must always account for relatedness, and we have seen that r provides the proper currency. In addition, there is no gain to benefit an individual who will not contribute to the future of the population. So net benefit must be valued by the degree of future contribution. That value is, by definition, v_p. Likewise, a cost to the actor must be rendered into a common currency measured by contribution to the future population. The proper exchange is v_a.

The problem with this extended rule is that the definitions are vague. Consider, for example, a population divided into age classes. Suppose the partner is a juvenile, with reproductive value v_p. This valuation accounts for several distinct processes. The juvenile's expected contribution to the future population depends on its survival to the age of reproduction, its fecundity as an adult, and when it reproduces.

But how should the net benefit, Bv_p, be measured? Suppose that the benefit reduces the amount of time required for the juvenile to mature into a reproductive adult, without changing net lifetime survival and fecundity. Then net benefit could be measured by the marginal change in the reproductive value of the future offspring born earlier as a result of the change in reproductive schedule. Here, it is not obvious how to separate the benefit, B, from the reproductive value weighting, v_p.

Real behaviors typically have multiple consequences. It is often difficult to account for all these consequences in terms of proper marginal costs and benefits and proper choice of reproductive value weightings. The only method available, if one begins with a rule such as Eq. (8.1), is to use intuition with no formal check.

Alternatively, the joint effects of kin selection and reproductive value can be accounted for within a standard procedure of maximization. I showed earlier that kin selection and coefficients of relatedness emerge from simple maximization methods (Section 4.3). Those methods begin with a function that expresses fitness. When the function is maximized with respect to changes in the effect of breeding value, the derivative (slope) of actor phenotype on recipient genotype emerges naturally as the relatedness coefficient for direct fitness.

The same approach works when recipients have different reproductive values. We begin with standard life history methods, which express the fitness function as a sum of fitness components, each weighted by reproductive value. Optimal trait values are found by maximizing fitness with respect to changes in the trait. Once again, direct fitness coefficients of relatedness emerge as derivatives (slopes) of actor phenotype on recipient genotype. Alternatively, we can use the slope of recipient genotype on actor genotype as the inclusive fitness coefficient of relatedness (Section 4.3).

To use this method it is first necessary to review life history theory. This theory provides the form of the fitness function needed for the maximization procedure.

REPRODUCTIVE VALUE OF EACH CLASS

Each individual must be valued according to its expected contribution to the future population, its reproductive value. One typically does not calculate reproductive value separately for each individual. Instead, individuals are separated into classes defined by some key attribute, such as sex, size, or age.

Three attributes of each class are important: number of individuals, reproductive value of each individual, and class reproductive value. A class j constitutes numerically a fraction u_j of the total population. The reproductive value of each class, c_j, is the fraction of all the genes in the distantly future population that come from individuals of class j. It follows that the reproductive value of each individual in class j is in proportion to

$$v_j = c_j/u_j. \tag{8.2}$$

It is important to distinguish the c_j from the v_j. For example, if the classes are the two sexes, male (m) and female (f), then in a population with haploid males and diploid females, a gene in the distant future has twice the probability of being in a female today as in a male today, so that $c_f = 2/3$ and $c_m = 1/3$ (Price 1970). By contrast, when we are working with age classes, we typically use the individual reproductive values, v_j, defined by Fisher (1958a, 27). In models of social behavior, the v_j are used as relative weights to compare fitness benefits to individuals of different classes.

The direct fitness method of maximization can be used in class-structured populations. First, as in Eq. (4.3), express total fitness in the population by

$$W = \sum c_j W_j \qquad (8.3)$$

where the fitness of each class, W_j, is weighted by its class reproductive value, c_j. Each W_j has the same value in a normal population, where *normal* denotes a population without genetic variation. The method is to perform standard direct fitness maximization with respect to transmitted breeding value through each fitness component, g_j', as

$$\frac{dW}{dg'} = \sum c_j \frac{dW_j}{dg_j'}.$$

If, for example, fitness is affected by characters y and z, then the derivative of each fitness component is expanded as

$$\frac{dW_j}{dg_j'} = \frac{\partial W_j}{\partial y}\frac{dy}{dg_j'} + \frac{\partial W_j}{\partial z}\frac{dz}{dg_j'},$$

replacing all derivatives (slopes) of characters on transmitted breeding values by coefficients of regression or relatedness.

The fitness components, W_j, must be defined properly. Suppose, for example, that the effect of a behavior on a class-j female is not to alter her overall fitness, but to modify different components of her fitness in different ways. If the recipient is a mother, her production of daughters may be affected differently from her production of sons, or perhaps her fecundity is affected but not her survival. We treat these components as different classes of offspring, and write the mother's fitness in terms of these components. For example, an age-2 mother who has five offspring surviving to next year and herself survives with probability s to age 3, would be regarded as dying and having five class-1 offspring and s class-3 offspring.

LIFE HISTORY: TECHNICAL DETAILS

I review standard life history theory in this section (Charlesworth 1994), using the notation in Taylor and Frank (1996). Let w_{ij} be the number of class-i offspring of a class-j individual. The count here is made according to genetic representation; that is, if an individual furnishes only one-half of the genes of an offspring, then that offspring is counted as one-half. The offspring matrix

$$\mathbf{A} = \left[w_{ij} \right]$$

records in the jth column the fitness components of a class j individual. The matrix $\mathbf{A}$ depends on variation in trait values. Define a normal population as one with no genetic variation and constant breeding value g^*, and normal fitness matrix $\mathbf{A}^*$. The dominant eigenvalue λ of $\mathbf{A}^*$ is the factor by which the normal population size is multiplied in each generation. The vector $\mathbf{v}$ of individual reproductive values is the dominant left eigenvector of $\mathbf{A}^*$,

$$\lambda \mathbf{v} = \mathbf{v} \mathbf{A}^*$$
$$\lambda v_j = \sum_i v_i w_{ij}^* \tag{8.4}$$

and the vector $\mathbf{u}$ of equilibrium class frequencies is the dominant right eigenvector of $\mathbf{A}^*$

$$\lambda \mathbf{u} = \mathbf{A}^* \mathbf{u}$$
$$\lambda u_i = \sum_j w_{ij}^* u_j.$$

The eigenvalue has a natural interpretation as a rate of increase. The stable distribution of class frequencies, $\mathbf{u}$, has elements summing to one. In a population of size n, the numbers in each class are $\mathbf{u}n$. One time period of growth is defined as one application of the fitness matrix $\mathbf{A}^*$; that is, the numbers in each class after one period are $\lambda \mathbf{u}n = \mathbf{A}^* \mathbf{u}n$. After t periods, by repeated application of the fitness matrix $\mathbf{A}^*$, the numbers in each class are $\lambda^t \mathbf{u}n$.

The fitness of a class-j individual is defined as the weighted sum

$$W_j = \sum_i \left(v_i / v_j \right) w_{ij}, \tag{8.5}$$

where the weight v_i / v_j is the relative reproductive value of a class-i offspring. Essentially the weights can be thought of as factors converting the fitness components into common units, which can then be added. A comparison of Eq. (8.4) and Eq. (8.5) shows that in a normal population, the W_j are all equal to λ. The conditions for Eq. (8.3) hold, so the total sum of fitness components with proper weights can be written as

$$W = \sum_j c_j W_j = \sum_{ij} v_i w_{ij} u_j = \mathbf{v} \mathbf{A} \mathbf{u}, \tag{8.6}$$

where the last term uses vector notation. Our goal is to study how a change in the transmitted breeding value, g', affects fitness, expressed as

$$\frac{dW}{dg'} = \sum_{ij} v_i \frac{dw_{ij}}{dg'_i} u_j = \mathbf{v} \frac{d\mathbf{A}}{dg'} \mathbf{u}. \tag{8.7}$$

This differentiation treats the v_i and u_j as constant, calculated from the normal g^* population, even though reproductive value and class frequency are often affected by the behavior. This is related to a standard result of life history optimization (Taylor and Frank 1996), as shown in the following paragraphs.

The standard measure of fitness for a rare mutant allele is not W but the dominant eigenvalue of λ of the matrix $\mathbf{A}$. This eigenvalue expresses the rate of growth of copies of the allele when the allele is in its equilibrium proportions among the classes.

The use of $dW/dg' = 0$ for maximization can be defended as follows. Write all terms as functions of g' to show their dependence on variants in (transmitted) breeding value. Then, by the definition of eigenvalues and eigenvectors

$$\mathbf{v}(g') [\mathbf{A}(g') - \lambda(g') \mathbf{I}] \mathbf{u}(g') = 0,$$

where $\mathbf{I}$ is the identity matrix. Denote differentiation with respect to g' by "~", and evaluate the derivatives at $g' = g^*$, yielding

$$\tilde{\mathbf{v}}[\mathbf{A}^* - \lambda\mathbf{I}] \mathbf{u} + \mathbf{v}[\tilde{\mathbf{A}} - \tilde{\lambda}\mathbf{I}] \mathbf{u} + \mathbf{v}[\mathbf{A}^* - \lambda\mathbf{I}] \tilde{\mathbf{u}} = 0.$$

Since $\mathbf{u}$ and $\mathbf{v}$ are eigenvectors, the first and last terms vanish, giving

$$\frac{dW}{dg'} = \mathbf{v}\tilde{\mathbf{A}}\mathbf{u} = (\mathbf{v} \cdot \mathbf{u})\, \tilde{\lambda},$$

as in Eq. (8.7), where everything is evaluated at $g' = g^*$. Since $\mathbf{v}$ and $\mathbf{u}$ are positive, $d\lambda/dg'$ and dW/dg' have the same sign. Thus, the derivative dW/dg' gives the correct direction of evolutionary change.

EXAMPLE ONE: DIRECT AND INCLUSIVE FITNESS

Evaluating dW/dg' automatically provides the proper coefficients of regression and relatedness. I illustrate these coefficients with a simple example (Taylor and Frank 1996). The purpose is to clarify various aspects of notation and method. I show the correspondence of the maximization procedure to standard inclusive fitness arguments, as in Eq. (8.1), and to the new direct fitness methods described in previous chapters. I present a more realistic example in the following section, to demonstrate the benefits of the formal maximization procedure over the commonly used inclusive fitness heuristic.

In this example there are three classes: (1) juvenile males, (2) juvenile females, and (3) adult females. Suppose the actors are the adult females, and all three classes are recipients. Each individual is affected by two phenotypes, y and z. For each individual, the trait y is the phenotype of the particular adult female with which the individual is associated. For juveniles, y is the mother's phenotype, and for an adult female, y is her own phenotype. The trait z is the average phenotype among an adult female's neighbors in the local group, excluding her own phenotype. If an adult female is in a group with her sisters, then from a juvenile male's point of view, z would be the phenotype of a randomly chosen aunt.

According to Eq. (8.3), the average direct fitness of recipients is

$$W = c_1 W_1 (y,z) + c_2 W_2 (y,z) + c_3 W_3 (y,z).$$

The direct fitness derivative is

$$\frac{dW}{dg'} = c_1 \frac{dW_1}{dg_1'} + c_2 \frac{dW_2}{dg_2'} + c_3 \frac{dW_3}{dg_3'}.$$

The derivatives on the right side can be expanded. For example, the derivative in the first term, for juvenile males, is

$$\frac{dW_1}{dg_1'} = \frac{\partial W_1}{\partial y} \frac{dy}{dg_1'} + \frac{\partial W_1}{\partial z} \frac{dz}{dg_1'}$$

$$= \frac{\partial W_1}{\partial y} \tilde{r}_1 + \frac{\partial W_1}{\partial z} \tilde{R}_1. \tag{8.8}$$

The direct fitness coefficients, $\tilde{r}$ and $\tilde{R}$, are the slopes of actor phenotype on recipient genotype. These coefficients were explained earlier (Chapter 4). For most problems, we could move quickly from this general expression for the derivative to a solution for the equilibrium phenotype favored by selection, as shown in the following section. My goal in the remainder of this section is to show the correspondence of the direct fitness form to the commonly used inclusive fitness expression.

Table 8.1 The Inclusive Fitness Effect of an Adult Female Actor

Recipients	Number	Effect	Relatedness	RV
Class 1				
Sons	n_1	a_1	r_1	v_1
Nephews	N_1	A_1	R_1	v_1
Class 2				
Daughters	n_2	a_2	r_2	v_2
Nieces	N_2	A_2	R_2	v_2
Class 3				
Self	1	a_3	1	v_3
Sisters	N_3	A_3	R_3	v_3

RV in the last column is reproductive value. The second column lists the numbers of recipients of each type for a single actor. The effects in the third column are the rates at which recipient fitness increases with changes in the actor phenotype, where fitness is normalized to unity. Thus, a_1 is the rate at which a male's fitness changes with respect to his mother's phenotype. The inclusive fitness effect, ΔW_{IF}, is the sum of the effects on different recipients:

$$\Delta W_{IF} = v_1 [n_1 a_1 r_1 + N_1 A_1 R_1] + v_2 [n_2 a_2 r_2 + N_2 A_2 R_2] + v_3 [a_3 + N_3 A_3 R_3].$$

The first step is to replace direct fitness coefficients by inclusive fitness relatedness coefficients. For example, we assume that $r_1 = \tilde{r}_1$, where the inclusive fitness coefficient r_1 is the slope of recipient genotype on actor genotype—in this case, the slope of the juvenile male's breeding value on the mother's breeding value. Use of this inclusive fitness coefficient requires that we include only that part of the direct fitness coefficient caused by shared breeding value and that the direction of the slope be reversed (Section 4.2).

For many problems, we can simply substitute the inclusive fitness coefficients for the direct fitness coefficients, because the problems concern a single phenotype with a common genetic basis in all individuals. But it is important to remember that the direct fitness coefficients are much more general. The inclusive fitness coefficients can be used only for a common, but restricted set of problems.

The remaining steps to obtain a classical accounting of inclusive fitness are tedious. I continue along to demonstrate the formal connection between the maximization method and the classical accounting shown in Table 8.1. I assume that each adult female is in a group with her sisters. This provides the convenient labels shown in Table 8.1 when the problem is examined from an adult female's point of view. She has sons and nephews among the juvenile males, daughters and nieces among the juvenile females, and herself and sisters among the adults. Any patterns of relatedness could be used. These labels simply make it easier to describe groups.

I continue expanding the first term for juvenile males in Eq. (8.8)

$$c_1 \left[\frac{\partial W_1}{\partial y} r_1 + \frac{\partial W_1}{\partial z} R_1 \right] = v_1 u_1 \left[k_1 a_1 r_1 + K_1 A_1 R_1 \right],$$

where k_1 and K_1 are the average number of mothers and aunts, respectively, of a juvenile male (e.g., k_1 is the probability that the mother of a juvenile male will be alive), and a_1 and A_1 are the rates at which a phenotypic change in a mother or an aunt affects juvenile male fitness. If we divide by u_3, the number of adult females, we get the first component of the inclusive fitness effect in Table 8.1. This follows because $(u_1/u_3)k_1 = n_1$, is the average number of sons that a mother affects, and $(u_1/u_3)K_1 = N_1$, is the average number of nephews affected by each aunt. To illustrate these last equations, if there were four juvenile males per adult female ($u_1/u_3 = 4$), and a male had on average three aunts, $K_1 = 3$, then an adult female would have on average $4 \times 3 = 12$ nephews. The full correspondence to Table 8.1 can be shown by continuing in this way for classes 2 and 3.

Example Two: Sex Ratio

The previous example is the typical sort of abstract problem used when discussing methods of analysis. The fitness effects are neatly summarized as parameters, so that the inclusive fitness condition is a simple accounting of weighted costs and benefits. In a real application the final answer must take the same form; that is, the properly weighted benefits must be greater than the weighted costs if the behavior is to spread. However, it is often quite difficult from the biology to see, in advance of the answer, what all of these effects are.

The maximization procedure has the advantage of leading from the biological assumptions to a proper answer, without having to guess in advance the form of the answer. I illustrate this distinction with a model of sex allocation that extends the example of Section 4.4. I discuss this problem at length in Chapter 10. Here my only goal is to study the technique, so I describe the problem briefly.

Consider a sexual population with N mated females breeding on each patch. Mating is random among the patch offspring, followed by partial dispersal of mated females at a rate μ, with cost c. Immigrant females compete for the N breeding sites with the nonmigrant females on each patch. This is the patch structure of the dispersal example in Section 7.2, in which we found the ESS dispersal rate. Here we want to calculate the ESS sex ratio (for a similar model, see Crespi and Taylor 1990).

Each mother splits her investment in offspring into fractions y sons and $1 - y$ daughters. An average neighbor of the female has phenotypes z and $1 - z$, where each female contributes $1/N$ to her patch's average phenotypes.

Adult females control phenotypes. The two recipient classes are the components of fitness of each adult female through male and female offspring. The female component is

$$W_f(y, z) = (1 - y)\left[(1 - \mu) p(z) + \mu(1 - c) p(\bar{z})\right],$$

where y is the female's mother's sex ratio, z is the average sex ratio on her native patch, and $\bar{z}$ is the average population-wide sex ratio. Here

$$p(z) = \frac{1}{N}\left[\frac{1}{(1 - z)(1 - \mu) + (1 - \bar{z})\mu(1 - c)}\right]$$

is the breeding probability of a female who competes on a z-patch. Note that the normalized value of p is

$$p(\bar{z}) = \frac{1}{N}\left[\frac{1}{(1 - \bar{z})(1 - c\mu)}\right].$$

It follows that the normal value of W_f is $1/N$. Similarly, the male fitness component is

$$W_m(y, z) = y\left(\frac{1 - z}{z}\right)\left[(1 - \mu) p(z) + \mu(1 - c) p(\bar{z})\right],$$

where the female : male mating ratio among offspring is $(1 - z)/z$. The normalized value of W_m is $1/N$, the same as the normal value of W_f.

These equations provide a full description of the biology. It is, in principle, possible to apply the inclusive fitness heuristic to obtain the ESS sex ratio. But most people would, I think, find it quite challenging to keep track of all the interactions and to combine them correctly.

Maximization allows the standard rules of differentiation to be applied in an automatic way. We begin with the average recipient fitness

$$W = c_m W_m + c_f W_f.$$

The equilibrium equation $dW/dg' = 0$, evaluated at $y = z = z^*$, yields the ESS sex ratio. The direct fitness regressions, $\tilde{r}$ and $\tilde{R}$ follow as usual from the slopes of actor phenotype on transmitted breeding value through each fitness component. We can, in many cases, replace the direct fitness coefficients by the inclusive fitness coefficients, r and R (see *Example: Sex Ratio*, p. 74). The ESS can be written as a ratio of number of males to females, $z^* : 1 - z^*$, yielding

$$c_m r_m - c_m R_m \; : \; c_f r_f + c_m R_m - k^2 \left(c_f R_f + c_m R_m \right), \qquad (8.9)$$

where $k = (1 - \mu)/(1 - c\mu)$ is the probability that a mated female is native to her breeding patch (Eq. (7.6)). The terms r_j are the inclusive fitness relatedness coefficients of a mother to her own sex-j offspring, and the R_j are the relatedness coefficients of the mother to a random patch offspring of sex j (including her own). One can interpret this result by realizing that it must follow the marginal theory of inclusive fitness, with additional reproductive value weights c_m and c_f. I take this up in Section 10.2.

SUMMARY OF MAXIMIZATION METHOD

The maximization method is, in practice, quite easy. Start with an expression for the fitness of each recipient class as a function of actor phenotypes. Weight each recipient class by its class reproductive value, maximize the weighted sum of recipient fitnesses with respect to variations in breeding value, replace slopes of recipient breeding values relative to actor phenotypes by appropriate relatedness coefficients, and solve for the equilibrium behavior.

Sometimes realistic problems work out neatly, such as the sex ratio examples. An application can, however, be quite challenging when different classes have multiple components of fitness. In this case, maximization of the demographic matrix, $\mathbf{A}$, provides a simple, formal procedure. The difficulties that arise are mainly technical. For example, a common problem focuses on a social group that contains various age classes, in which the behavior of each class affects the fitness of relatives in other classes. If each age class potentially takes on different values for a trait, then we must consider the simultaneous optimization of many different kinds of behaviors, each influencing several different classes of recipients.

Simultaneous optimization does not, in principle, create special difficulties. But there are two problems. First, this type of optimization can be technically difficult. Second, the methods outlined above are still rather new, and there are few good examples of applications.

In short, the subject of life history in social groups has not been developed very well. The basic outline of the techniques seems to be in place, but other issues are likely to arise in application. The following sections are a collection of preliminary analyses. They are constructed to highlight interesting conceptual and technical problems, and to suggest a few applications that could be developed.

8.2 Child Mortality in Social Groups

Fisher (1958a) noted that age-specific reproductive values of humans increase up to the age of first reproduction, and then decline throughout adult life. The lower reproductive value of children is explained by the fact that a child must first survive to reproductive age before it contributes to future generations. The declining reproductive value of adults is explained by the fact that age-specific survival and fecundity decline with age. An older adult therefore has a lower expectation of future reproduction than a younger adult (Fig. 8.1).

The human mortality curve follows inversely the trend of the reproductive value curve (Fig. 8.1). Periods of high mortality occur early and late in life, when reproductive value is low. Minimum mortality occurs just before the age of peak reproductive value.

Fisher suggested that the shape of the mortality curve could be explained by the reproductive value curve. The problem of age-specific mortality (senescence) has subsequently received much attention (Rose 1991; Charlesworth 1994). I focus here on the high level of juvenile mortality observed in some species.

Hamilton (1966) emphasized that a simple model of natural selection never predicts higher mortality early in life. The essence of his argument can be written as follows. Let fitness, w, be the expected number of

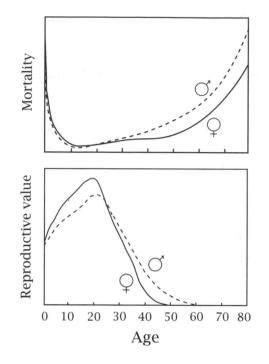

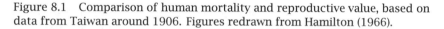

Figure 8.1 Comparison of human mortality and reproductive value, based on data from Taiwan around 1906. Figures redrawn from Hamilton (1966).

offspring produced in a lifetime, which is the sum of the fecundity, f_i, at each age, i, weighted by the probability, p_i, that the organism lives to age i

$$w = \sum_{i=0}^{\infty} f_i p_i = \sum_{i=0}^{\infty} f_i \prod_{j=0}^{i} s_j,$$

where s_j is the probability of survival during the jth age period.

To study the effect on fitness of a change in survival at a particular age, k, let $\hat{s}_k = s_k(1 + \delta)$, yielding the new fitness

$$\hat{w} = \sum_{i=0}^{k-1} f_i \prod_{j=0}^{i} s_j + (1 + \delta) \sum_{i=k}^{\infty} f_i \prod_{j=0}^{i} s_j.$$

The change in fitness for a change in survival at age k is

$$\hat{w} - w = \delta \sum_{i=k}^{\infty} f_i \prod_{j=0}^{i} s_j.$$

It is clear that a change at an earlier age (smaller k) always has an effect that is greater than or equal to a change imposed at a later age.

This simple theory predicts that age-specific mortality should be a nondecreasing function of age. But mortality actually decreases sharply from birth to the early teens in humans. Hamilton (1966) suggested various kin selection processes to explain this discrepancy. Perhaps the simplest scenario, not mentioned by Hamilton, is to imagine a stressful event in the family, such as a temporary shortage of food. In a severe case, where some offspring can be protected and others must be put at risk of mortality, what is the optimal allocation of resources by a parent? The answer is simply to favor those offspring with higher reproductive value. Such a process would likely lead to a mortality curve that is approximately the inverse of the reproductive value curve.

The parental control explanation is not satisfactory, however, because the mortality rates seem to be intrinsic to the offspring under uniformly good conditions. Thus Hamilton mentioned how selection might act directly on the offspring to increase mortality and enhance the survival of siblings or parents. I use this problem to motivate some models of reproductive value and kin selection. My main purpose is to consider conceptual and technical issues, rather than the problem of offspring mortality.

Simplest Models

I begin with a simple model of family altruism (extending Taylor and Frank 1996). Let a population consist of progeny (class p) and adults (class a). Individuals are hermaphroditic, reproducing as both male and female. Adults mate in each year, and the resulting progeny become juveniles in the following year. Each adult expects $2n$ offspring to survive to the juvenile stage, n of its own (it is the "mother" of these) and n through mating with other adults. An adult is credited with one-half of each offspring for its gametic contribution. Thus each adult receives credit for $n = 2n/2$ offspring. Juveniles do not reproduce, and they survive to the next year with probability s. Adults of any age survive to the following year with probability t.

The two classes form a normal fitness matrix

$$\mathbf{A}^* = \begin{bmatrix} 0 & n \\ s & t \end{bmatrix}.$$

The dominant eigenvalue λ of A^* is the larger root of the characteristic equation

$$\lambda^2 - t\lambda - sn = 0.$$

Following standard life history theory, the present value of all juvenile offspring of a current juvenile is set to unity

$$1 = \lambda^{-2}sn + \lambda^{-3}stn + \lambda^{-4}st^2n + \ldots = \frac{sn}{\lambda(\lambda - t)}. \tag{8.10}$$

Here λ is the time discount rate, such that the λ^{-2} accounts for the fact that a juvenile's first set of offspring will be juveniles two years into the future. At that time, the population will have changed size by a factor λ^2; thus the proportion of the total population represented by those future juveniles is weighted by λ^{-2}.

Standard calculations yield the stable class frequencies and reproductive values

$$\mathbf{u} \propto [n \quad \lambda]$$
$$\mathbf{v} \propto [s \quad \lambda],$$

where $\mathbf{u}$ is a column vector, and $\mathbf{v}$ is a row vector. The symbol $\propto$ means *proportional to;* that is, multiplicative constants can be chosen arbitrarily. If one follows Fisher's (1958a) normalization of reproductive value as the total number of expected offspring discounted to present value, then $v_p = 1$, as in the sum in the middle of Eq. (8.10), and adult value is $v_a = \lambda/s$.

This demographic analysis focuses on whole individuals—progeny and adults. Later it will be necessary to consider the success of the hermaphroditic adults separately via male and female fitness components. In this case we need to follow gametic success in the formation of progeny, with normal matrix

$$A^* = \begin{bmatrix} 0 & 0 & n/2 \\ 0 & 0 & n/2 \\ s & s & t \end{bmatrix}, \tag{8.11}$$

where columns and rows are, from top to bottom and left to right, male gametes, female gametes, and adults. For example, the n progeny produced by an individual via its mating success as a male are weighted by $1/2$ to account for gametic contribution to each offspring. This matrix has a dominant eigenvalue described by the same characteristic equation as the whole individual matrix. The stable class frequencies and reproductive values are

$$\mathbf{u} \propto [n/2 \quad n/2 \quad \lambda]$$
$$\mathbf{v} \propto [s \quad s \quad \lambda].$$

The role of the various components is best shown by three examples. In the first, the behavior affects all fitness components in the same way, so detailed separation by components is unnecessary. The second model assumes that the behavior affects fitness components differently, so it is necessary to pay attention to class-specific properties in the whole individual matrix. In the third model, male and female components of fitness are affected in different ways; thus the analysis requires the full gametic matrix.

MODEL 1. ALL FITNESS COMPONENTS AFFECTED THE SAME WAY

The problem here is altruistic behavior of progeny toward their adult mother, given that the mother has survived. The average fitness of recipients is

$$W = \sum c_j W_j = u_p v_p W_p + u_a v_a W_a \propto ns W_p + \lambda^2 W_a, \qquad (8.12)$$

where, in this case, I use the 2×2 whole individual matrix for progeny and adults.

Suppose that a single altruistic interaction increases the mother's fitness by a factor of b but decreases the juvenile's fitness by a factor of c. Let x be the phenotype of a juvenile, and let y be the average phenotype of a mother's juvenile offspring. The fitnesses of the two recipient classes, normalized to unity for $x = y = 0$, are

$$W_p = 1 - tcx \qquad (8.13)$$
$$W_a = 1 + (t/\lambda) nby, \qquad (8.14)$$

where t in Eq. (8.13) is the probability that the mother of the juvenile is alive, and t/λ in Eq. (8.14) is the probability that a random adult was an adult of the previous year. An adult who bred in the previous year can expect ny altruistic interactions from her offspring. Differentiating fitness, W, as given in Eq. (8.12)

$$\frac{dW}{dg'} = nt \left[-sc \frac{dx}{dg'_p} + \lambda b \frac{dy}{dg'_a} \right].$$

The derivative dx/dg_p' is the slope of a progeny's phenotype on its transmitted breeding value. A progeny transmits to the next time period by surviving, thus its transmitted and current breeding values are the same, $g_p' = g_p$, and $dx/dg_p' = 1$. Let $\tilde{r} = dy/dg_a'$ be the slope of a randomly chosen juvenile's phenotype, y, on its mother's transmitted breeding value, g_a'. I will clarify the interpretation of $\tilde{r}$ and the assumptions of this analysis in model 3 below. For now, I simply note that under random mating, the direct fitness coefficient $\tilde{r}$ can be replaced by the inclusive fitness coefficient r, in which r measures the juvenile's (actor's) relatedness to its mother.

The behavior is favored when $dW/dg' > 0$, which gives

$$sc < \lambda br. \qquad (8.15)$$

Offspring are favored to increase their mortality if that increase provides sufficient benefit to mothers. The offspring suffers a cost, c, for the act, and the mother gains b. This $c : b$ ratio must be weighted by individual reproductive values, $s : \lambda$, and by the ratio of relatedness coefficients, $1 : r$, taken from the actor's (offspring's) point of view. Thus this result can be obtained by a standard application of Hamilton's rule, simply adding the appropriate reproductive value weightings. Charlesworth and Charnov (1981) were perhaps the first to construct a formal model of kin selection with reproductive value weightings in an age-structured population.

MODEL 2. DIFFERENT FITNESS COMPONENTS AFFECTED DIFFERENTLY

Suppose that an altruistic behavior by the juvenile increases the mother's current fecundity by a factor of b but does not affect her survival. Because the fitness components are affected differently, we must work directly with the fitness matrix

$$\mathbf{A} = \begin{bmatrix} 0 & n\,(1 + tnby/\lambda) \\ s\,(1 - tcx) & t \end{bmatrix},$$

where, in this case, we are still using the whole individual matrix. Differentiating as in Eq. (8.7)

$$\frac{dW}{dg'} = \mathbf{v}\frac{d\mathbf{A}}{dg'}\mathbf{u} = [s \quad \lambda] \begin{bmatrix} 0 & n^2 tb\tilde{r}/\lambda \\ -stc & 0 \end{bmatrix} \begin{bmatrix} n \\ \lambda \end{bmatrix}$$

$$= nst\,[-\lambda c + bn\tilde{r}],$$

Table 8.2   Inclusive Fitness Calculation for a Model of Juvenile Mortality

Method 1: count "offspring" in the next year for each recipient			
Recipient	Offspring	RV	Relatedness
Mother	$2nb$	s	$r/2$
Juvenile	$-sc$	λ	1
$\Delta W_{IF} = nbsr - sc\lambda = s(nbr - c\lambda)$			

Method 2: measure fitness effects in terms of juveniles			
Recipient	Juvenile offspring	Time discount	Relatedness
Mother	$2nb$	λ	$r/2$
Juvenile	$-c$	1	1
$\Delta W_{IF} = nbr/\lambda - c = (nbr - c\lambda)/\lambda$			

There are two different ways one might construct the inclusive fitness argument leading to Eq. (8.16). Method 1 follows the matrix approach of the text. The two components of fitness are treated as next year's "offspring," and the class of the offspring must be noted and weighted by the class reproductive value (RV). However, all these "offspring" belong to next year, so no weighting by time is required. Method 2 measures all fitness effects as "juvenile" units, either this year or next year. Then RV weights are not required, but next year's juveniles must be discounted by the population growth factor, λ. Note that the mother's future offspring have relatedness to the actor of $r/2$, assuming the mother's mate is unrelated to the actor. This conforms to the assumptions that r is the relatedness of the juvenile actors to their mother, and that mothers remate randomly in each year.

where, as before, $dx/dg'_p = 1$ and $dy/dg'_a = r$, in which r is the relatedness of juveniles to their hermaphroditic mother. The behavior is favored when $dW/dg' > 0$ is positive, yielding the condition

$$c < \frac{bnr}{\lambda}. \tag{8.16}$$

Two different ways to formulate a modified Hamilton's rule are shown in Table 8.2. The key is that a current offspring values an increase in its mother's fecundity by comparing its own reproductive value, s, relative to the reproductive value of next year's offspring, s/λ. Here $1/\lambda$ is the time discount for future offspring (Rogers 1993; Taylor and Frank 1996). In a stable population, the time discount is $\lambda = 1$, and the weighting vanishes.

MODEL 3. FULL GAMETIC MODEL

Nonrandom mating affects in different ways a hermaphroditic mother's male mating success, female fecundity, and survival. This requires attention to the full gametic matrix in Eq. (8.11). Expanding the matrix to include fitness effects of juvenile altruism, we obtain

$$\mathbf{A} = \begin{bmatrix} 0 & 0 & n\beta\,(b_m, y)\,B\,(b_f, z)\,/2 \\ 0 & 0 & nB\,(b_f, y)\,/2 \\ sC\,(x) & sC\,(x) & tB\,(b_s, y) \end{bmatrix}, \qquad (8.17)$$

where the beneficial effect of juveniles with phenotype y on fecundity or survival of their parents is

$$B\,(\alpha, y) = 1 + (t/\lambda)\,n\alpha y,$$

and the beneficial effect of the juveniles on the mating success of the male attached to their mother is

$$\beta\,(\alpha, y) = \frac{1 + (t/\lambda)\,n\alpha y}{1 + (t/\lambda)\,n\alpha \overline{y}}.$$

The mating success term is normalized by the average phenotype, $\overline{y}$, in the local population in which male gametes compete. This constrains the average number of mates to one. This constraint is imposed because, in hermaphroditic populations, there is always one female per male. Finally, the cost of altruism to a juvenile with phenotype x is

$$C\,(x) = 1 - tcx,$$

as in the prior models.

The upper-right term in the matrix, $n\beta B/2$, is the success of male gametes. The function β gives the increase in the expected number of matings. Mating success benefits from altruism of the juveniles produced by the attached mother, an outcome of hermaphroditism. The mates of the male have juveniles with phenotype z, which increases the fecundity of the male's mates by $B(b_f, z)$. The second row in the last column is the success of female gametes of the mother. The lower-right entry is the survival of the mother and the male gametes that she receives from her mates.

The condition for an initial increase in juvenile altruism is obtained in the usual way

$$\frac{dW}{dg'} = \mathbf{v}\frac{d\mathbf{A}}{g'}\mathbf{u} > 0,$$

evaluated at $x^* = y^* = z^* = 0$, yielding

$$nsb_f \, (\tilde{r}_f + \tilde{r}_\mu) \, /2 + t\lambda b_s \, (\tilde{r}_f + \tilde{r}_m + \tilde{r}_\mu) \, /2 + nsb_m \left(\tilde{r}_m - \tilde{R}_m \right) /2 > s\lambda c.$$

$$(8.18)$$

The three terms on the left represent the effects of the juveniles on the fecundity, survival, and male mating success of the mother.

Fecundity

The relatedness coefficient, $\tilde{r}_f = dy/g'_f$, is the slope of the juveniles' average phenotype in a brood, y, relative to the breeding value of the mother's transmitted female gametes. The coefficient $\tilde{r}_\mu$ is dz/dg'_m. The term z is the phenotype of juveniles produced by the mates of a male. This phenotype influences the transmission of the male's gametes, g'_m, to next year's brood.

The combined relatedness coefficient weighting for the female fecundity component is $(\tilde{r}_f + \tilde{r}_\mu)/2$. When a juvenile actor influences the female fecundity of its mother, the proper direct fitness coefficient uses the mother's transmitted breeding value as the recipient. The mother's transmitted breeding value is the average of her own breeding value and the breeding value of her mate. This is equivalent to the breeding value of the mother's future progeny produced as a female.

Survival

Increased survival of the mother affects the transmission of three gametic components. Survival carries forward the female and male gametes of the mother to the next generation. The juvenile's relatedness to female gametes is $\tilde{r}_f$, given above, and its relatedness to male gametes is $\tilde{r}_m = dy/dg'_m$. The mother's male and female gametes have the same breeding value, thus the relatedness between juvenile and mother is $\tilde{r} = \tilde{r}_f = \tilde{r}_m$.

The surviving mother will also transmit the genes of her mates when she reproduces as a female. The relatedness between juveniles and mother's future mates was defined above as $\tilde{r}_\mu$. This term is weighted by one-half because it is a gametic effect. Put another way, the juveniles influence the male fecundity of their mother's mates, but not the female fecundity of the mates.

Male mating success

The term $\tilde{r}_m = dy/dg'_m$ is the slope of offspring phenotype on the transmitted breeding value through male gametes of the mother. The coefficient $\tilde{r}_m/2$ is the relatedness between a juvenile and the half of future offspring produced by male gametes of the juvenile's mother. The one-half arises because the juvenile influences only the male parent's mating success, not the fecundity of the male's mates.

An increase in the average progeny phenotype in the local breeding group, $\bar{y}$, raises the number of competing male gametes. This reduces the success of male gametes transmitted by an adult. The relatedness of a random offspring in the local breeding group to a male gamete is $\tilde{R}_m = d\bar{y}/dg'_m$. This term arises as a negative factor because it reduces the success of relatives. Division by one-half is for the same reason given in the prior paragraph.

INCLUSIVE FITNESS COEFFICIENTS

The direct fitness coefficients can be replaced by inclusive fitness coefficients. The condition in Eq. (8.18) can be rewritten, using inclusive fitness coefficients, as

$$nsb_f \left(r + r_\mu \right)/2 + t\lambda b_s \left(r + r_\mu/2 \right) + nsb_m \left(r - R_m \right)/2 > s\lambda c, \quad (8.19)$$

where coefficients are taken from the juvenile actor's point of view. The coefficient r is the juvenile's relatedness to its mother, r_μ is the juvenile's relatedness to its mother's mates, and R_m is the juvenile's relatedness to neighboring hermaphrodites that contribute male gametes to the local breeding pool.

COMPARISON OF MODELS

The results of models 1 and 2 can be obtained directly from Eq. (8.19). This exercise illustrates the formal relations among the models and how the different methods should be interpreted.

The first model assumed that all fitness components of the mother were influenced in the same way. Thus the third column of the matrx in Eq. (8.17) must be affected by the same fitness term. If we assume that the benefit coefficients are the same, $b_i = b$, and mating is random so that $r_\mu = R_m = 0$, then the general condition in Eq. (8.19) reduces to the conclusion from model 1 in Eq. (8.15). Demonstrating equivalence requires one to use the identity $\lambda^2 = \lambda t + sn$ from the characteristic equation.

For model 2, the only effects are on the fecundity and mating success, $b_f = b_m = b$, and $b_s = 0$. The result in Eq. (8.16) follows.

Finally, the general model in Eq. (8.19) can be reduced in a variety of ways. For example, when the benefits are all equal, $b_i = b$, then

$$b\left[\lambda\left(r + r_\mu/2\right) - n\left(s/\lambda\right)R_m/2\right] > sc.$$

The term $\lambda(r + r_\mu/2)$ is the product of the reproductive value of the mother, λ, and the juvenile's relatedness to mother's transmitted breeding value through all her fitness components, survival, fecundity, and male mating success. The mating competition discount is $n(s/\lambda)R_m/2$. The number of progeny through mating is n, the reproductive value of next year's offspring is s/λ, and the juvenile's relatedness to male gametes that compete with its mother's male gametes is R_m. The one-half arises because the juvenile influences only the male parent's mating success, not the fecundity of the male's mates.

CRITIQUE

These models clarify the forces that act on altruistic behavior. But they may be too simple, as is often the case in formulations that end up in Hamilton-rule form. For example, Eq. (8.15) suggests that when survival of offspring, s, is already low, then offspring are favored to reduce their survival more, in order to aid their mothers. As offspring survival declines, there would seem to be an accelerating benefit from even lower survival. Surely survival will not be favored to decrease to zero. A similar lack of clarity occurs in Eq. (8.16). If the condition is satisfied, then offspring are favored to decrease their survival in order to increase the fecundity of their mothers. But when does the decrease in survival stop?

The problem with these formulations is that, in simplifying so much, the marginal changes in cost and benefit are hidden within the model. There is, of course, nothing wrong with these formulations. But almost the entire literature of social evolution is formulated in terms of these vague inequalities, without any hint about how to extract more meaningful statements. When explicit statements about comparative statics are desired, authors often turn to a variety of ad hoc methods or computer simulations. The methods outlined above suggest that a slightly modified procedure of standard maximization may be the right approach. I explore some problems and examples in the following sections.

POPULATION GROWTH: VARIABLE OR PARAMETER?

I described in the previous section the standard optimization technique for life history analysis, with modifications to handle social interactions. That method has the virtue of a clear formalism, with techniques to translate survival and fecundity schedules into reproductive values, $\mathbf{v}$, and the rate of population growth, λ. Thus all demographic assumptions and consequences are fully integrated into the formulation and analysis.

A result with the full life history analysis, such as Eq. (8.16), may seem to take population growth, λ, as a fixed parameter. But that problem was constructed with the normal value of the character, $x = y = y^* = 0$; thus one can only examine the direction of evolutionary change. As the normal value of y^* changes from zero, survival and fecundity change, and hence λ changes. That analysis could be extended by allowing normal character value $y^* \neq 0$, adjusting normalizations for class fitness, and making explicit the dependence of λ on y^*. The extended analysis would be messier, but easy enough to complete.

The difficulty with an extended analysis is that survival and fecundity are likely to be influenced by forces other than social interaction. One would have to make such density-dependent factors explicit, further complicating the details of the model. But one may then lose the original goal of focusing simply on the evolution of a social character. Alternatively, one can take population growth, λ, as an extrinsic parameter. This assumption is valid if extrinsic forces regulate population growth and those forces do not influence the marginal costs and benefits of the social character under study. An example follows.

CYCLE FITNESS

The net reproductive rate of a female is commonly used as a measure of fitness in complex demographies (Charlesworth 1994). This measure, often labeled R_0, takes on various forms depending on the biology. For example, Charnov (1993) makes extensive use of

$$R_0 = S(\alpha) F(\alpha),$$

where $S(\alpha)$ is survival to the age of first reproduction, α, and $F(\alpha)$ is the expected lifetime fecundity of an individual who survived to age α. In the study of parasite life history, S is often taken as the expected time a parasite survives within a host, and F is the number of new hosts that parasite infects per time unit (Anderson and May 1991). Caswell (1985) developed a measure of fitness over complex life cycles, which is similar to the approach I take here.

For social behavior, I prefer to use my standard notation of w for the expected value of individual fitness, and measure w over a life cycle. For example, fitness for the model leading to Eq. (8.16) can be written as

$$w = s\,(1 - tcx)\left(nt/\lambda + n\,(t/\lambda)^2\,(1 + nby) + n\,(t/\lambda)^3\,(1 + nby) + \ldots\right),$$

where $s(1 - tcx)$ is the probability of survival to the adult stage. The first tn/λ in the sum is the probability, t, of survival during the first adult season, in which case n offspring are produced and discounted by the population growth rate, λ. The probability of survival to the second adult season is t^2, the discount is λ^2, and fecundity is n multiplied by $1 + bny$, the benefit of altruism from the n juveniles of the prior year. Remaining terms are calculated in a similar way. If we use the standard simplification for geometric series, this fitness is

$$w\,(x, y) = (1 - tcx)\,(1 + nbyt/\lambda)\left(\frac{stn}{\lambda - t}\right). \qquad (8.20)$$

This function tracks the fitness of a single individual who expresses phenotype x as a juvenile and later, as an adult, interacts with juveniles who have average phenotype y. Thus the individual is an actor when a juvenile and is a recipient when a juvenile and when an adult.

By our standard method, we evaluate $dw/dg' = 0$ at $x = y = y^*$. The direct fitness coefficient, $\tilde{r} = dy/dg'$, is replaced by the inclusive fitness coefficient, r, the slope of mother's (recipient's) transmitted breeding value on an average juvenile's (actor's) breeding value. I assume mating is random; thus r is the juvenile's relatedness to its mother. This yields the solution

$$y^* = \frac{nbr - c\lambda}{cnbt\,(1 + r)}. \qquad (8.21)$$

Note that the numerator presents the same condition as Eq. (8.16) for $y^* > 0$, but we now have the scaling effect of various parameters in the denominator. This provides the full information needed to calculate the equilibrium if we set λ as a parameter. The usual approach for this is to set the normal value of the net reproductive rate at $w(y^*, y^*) = 1$, which implies $\lambda = 1$. This constraint must be absorbed by adjusting one of the remaining parameters. The proper choice depends on the biology. A common assumption is to impose density-dependent mortality on juvenile survivorship, s, by expressing s in terms of the other parameters to satisfy $w = 1$ (e.g., Charnov 1993, assuming the other parameters have magnitudes that allow this expression). This form of density dependence has no effect on the equilibrium trait value in Eq. (8.21), which is independent of juvenile survivorship.

MATERNAL CONTROL

What if the mother controls the phenotype of the juveniles in her group? The fitness function is now

$$w(y, x) = (1 - tcy)(1 + nbxt/\lambda)\left(\frac{stn}{\lambda - t}\right),$$

where the only difference from Eq. (8.20) is that x and y have been switched. This switch occurs because the offspring behave according to the expected phenotype of their mother, y, and a mother has offspring that behave according to her own phenotype, x. The equilibrium is

$$y^* = \frac{nb - rc\lambda}{cnbt(1 + r)},$$

where r is the inclusive fitness relatedness of the mother to her progeny, that is, the slope of juveniles (recipients) on mothers (actors). As r declines, the mother causes the juveniles to sacrifice more, reducing their survival and increasing her fecundity. For $r < 1$, the mother causes the juveniles to sacrifice more than they would of their own accord.

ACTORS IN MORE THAN ONE CLASS

The simple kin selection models show how increased offspring mortality can be favored. Those models assume a single juvenile age class. How do mortality schedules evolve when there are several juvenile age classes? The comparative statics problem is to find the optimum conditional behavior for each class, given that all other classes are also behaving optimally.

This is a standard type of problem in life history (Charlesworth 1994) and behavior (Oster and Wilson 1978; Mangel and Clark 1988). Simultaneous optimization is conceptually simple, but technically difficult. I avoid the technical issues here, which are typical aspects of applied mathematics. My goal is to show, once again, that kin selection can be studied easily in such optimization problems. The extended kin selection method is important, because nearly all realistic models of social groups require analysis of different classes. Each class has the potential to adjust its own behavior and to influence the fitness of other classes within the group.

I assume a stable population, so that we can use cycle fitness. As before, the fitness of a life cycle can be divided as

$$w = S(m) F(m),$$

where $S(m)$ is survival to age of maturity, m, and $F(m)$ is expected lifetime fecundity of a mature individual of age m. The juveniles are labeled by age, $0, 1, \ldots, m$. The socially independent survival of each individual of age i is τy_i. Our goal is to find the optimal set $\{y_i^*\}$. The average survival of each age class within the social group, ignoring social interactions, is τz_i. The individual and average group phenotypes can be written in vector notation as $\mathbf{y}$ and $\mathbf{z}$, respectively, with fitness as $w(\mathbf{y}, \mathbf{z})$ to emphasize the nature of the optimization problem.

Survival through age i is the product of survivals at each age k

$$S(i) = \prod_{k=0}^{i} \tau y_k \left(1 - \sum_{j=0}^{m} a_{kj} n_j \right), \tag{8.22}$$

where n_j is the number of juveniles of age j, and a_{kj} is the effect of a juvenile of age j on the survival of an individual of age k. I assume $a_{kj} > 0$, so that each juvenile has a negative effect on the survival of all other juveniles, a form of density-dependent competition within the social group. If the older juvenile classes were helpers, then $a < 0$, and those helpers would increase the survival of other juvenile classes. The formulation is therefore a general description for social interactions within a complex group. To simplify the problem, I have assumed that each new social group begins its cycle with a stable age distribution of juveniles. It would be more realistic to follow the development of each social group from the production of its first brood through the end of its cycle.

The number of juveniles in the group of age j is

$$n_j = n\overline{S}(j),$$

the number of juveniles born in each period, n, multiplied by $\overline{S}(j)$, average survivorship to age j. The total number of juveniles of all age classes is

$$\pi = n \sum_{j=0}^{m} \overline{S}(j).$$

The average group survivorship, $\overline{S}(j)$, depends on average phenotypes, z, and can be obtained from Eq. (8.22), replacing y_k with z_k. The expected lifetime fecundity of a mature female of age m is

$$F(m) = n \sum_{i=0}^{\infty} \left[y \left(1 - \sum_{j=0}^{m} b_j n_j \right) \right]^i = \frac{n}{1 - y \left(1 - \sum_{j=0}^{m} b_j n_j \right)},$$

where b_j is the effect of juveniles of age j on the survival of their mother, and the right term is obtained by the standard geometric series simplification.

The full expression for cycle fitness is therefore

$$w(\mathbf{y}, \mathbf{z}) = \frac{n \tau^{m+1} \prod_{k=0}^{m} y_k \left(1 - \sum_{j=0}^{m} a_{kj} n_j \right)}{1 - y \left(1 - \sum_{j=0}^{m} b_j n_j \right)},$$

where the n_j are functions of z.

My purpose is to show the structure of the problem rather than to analyze all possible solutions. To simplify the analysis, I assume that juveniles do not influence the survival of their mother, $b_j = 0$ for all j, and that each age class has the same effect on other age classes, $a_{kj} = a$. The fitness function can then be written, dropping constants, as

$$w(\mathbf{y}, \mathbf{z}) = \left(\prod_{k=0}^{m} y_k \right) (1 - a\pi)^{m+1}.$$

Maximizing $\log(w)$ is equivalent to maximizing w because $\log(w)$ increases with w. The form of $\log(w)$ is

$$\log[w(\mathbf{y}, \mathbf{z})] = \left(\sum_{k=0}^{m} \log(y_k) \right) + (m+1) \log(1 - a\pi).$$

The problem is to find the set of equilibrium trait values, $\mathbf{y}^*$, such that for each age class, i, small variants in the transmitted breeding value, g_i', for the trait, y_i, cause lower fitness. I assume throughout that individual and transmitted breeding values are equal, $g_i = g_i'$.

The optimum must occur within the region $0 \leq y_i \leq 1$ for each i. Finding a local maximum in this region is not always easy because the maximum may occur on the boundary, where at least one y_i is zero or one. Thus, solving the simultaneous set of equations, $d\log(w)/dg_i = 0$ for $i = 0, 1, \ldots, m$, will often fail to yield a candidate equilibrium within the search region. Nonetheless, we can often learn something about the system by examining the derivatives. In this case, for each age i, we are interested in how fitness changes with a small change in breeding value, g_i, evaluated at a point without allelic variation, $\mathbf{z} = \mathbf{y}$

$$\frac{d\left[\log(w)\right]}{dg_i} = \frac{1}{y_i}\frac{dy_i}{dg_i} - \frac{\alpha(m+1)\pi_i}{(1-\alpha\pi)}\frac{dz_i}{dg_i},$$

where $\pi_i = \partial\pi/\partial z_i$. Note that π_i decreases as i increases, because a rise in survivorship at a later age has a smaller effect on the sum of average survivorships over all ages. It may be that, in a social group, the total number of juveniles, π, is regulated to a constant by density effects. The term π_i is really a measure of "partial or instantaneous pressure" caused by a change in the ith age class, rather than a measure of net change in survivorship and juvenile numbers. This structure of selective forces is typical of density-dependent selection (Charlesworth 1994).

As usual, we normalize g_i so that $dy_i/dg_i = 1$ and thus $dz_i/dg_i = \tilde{r}_i$. The derivative can now be written as

$$\frac{d\left[\log(w)\right]}{dg_i} = \frac{1}{y_i} - \frac{\tilde{r}_i a(m+1)\pi_i}{(1-\alpha\pi)}. \tag{8.23}$$

Differentiation allows three simple conclusions. First, if $\tilde{r}_i = 0$, then the derivative is always positive, so that selection always favors $y_i \to 1$. In words, if an act affects only unrelated individuals, then survival is favored to increase to a maximum. However, $\tilde{r}_i = 0$ is an unlikely case because, in this model, an actor is also part of the group of recipients, so with n unrelated individuals, $\tilde{r}_i = 1/n$, showing the contribution of the actor to the group average.

The second interesting conclusion from differentiation is that only relatedness among individuals of the same age class matters. It is irrelevant that the fitnesses of recipients in other age classes, who may be related to the actor, are strongly affected by the actor's behavior.

The reason is that survival in each age class, y_i, is assumed to be a separate trait uncorrelated with survival in other age classes, y_j; that is, $dy_j/dg_i = dz_j/dg_i = 0$. The way in which correlated characters affect simultaneous selection on multiple traits was discussed in Section 6.2.

The third point is that selection favors a reduction in survival more strongly as age decreases. This can be seen by noting that in Eq. (8.23), the derivative declines with decreasing age, i, because π_i increases as i becomes smaller. This makes sense because reduced survival at a particular age provides survival benefits to relatives in the cohort at all later ages. Thus, if selection favors reduced survival in any age class, then survival in the youngest age class must be reduced such that $y_0 < 1$. Indeed, it seems plausible that survival would be reduced only in the earliest age class—a single demographic adjustment that would increase colony throughput at later ages. However, overall colony throughput would likely be maximized only when $\tilde{r} = 1$.

To sum up, this section provided tools to study the life history of social groups. It is not easy, without a formal approach, to understand the many complex selective forces in social insects, cooperatively breeding birds, lions, chimpanzees, or humans. Although I focused on a simple model of mortality, one can readily extend the analysis to both harmful and beneficial social acts by different classes. The classes themselves can be defined by variables other than age, such as condition or caste.

This mortality model suggests ways in which to expand the analytical tools and concepts of social evolution. For example, Hamilton (1966) mentioned that earlier age classes might have elevated mortality thresholds. The lives of the young would give out when the combination of condition and age drops the juvenile's expected reproductive value below that of the reproductive value of a replacement sibling, weighted by relatedness to the later sibling. This suggests an expanded model of conditional behavior, combining the tools outlined in Section 6.2 with the class structure of this section.

Juvenile mortality may also be influenced by risk control in the face of uncertain conditions (Godfray et al. 1991). More juveniles may be started out in life than, on average, can be raised. If conditions prove to be harsh, then age-structured mortality thins the family by concentrating the mortality on the young. If resources are plentiful, then more young than average could be raised. Perhaps stochastic dynamic programming (Mangel and Clark 1988), with extensions to handle reproductive value and kin selection, could be developed to analyze this type of problem. Other tools from economic analysis may also prove valuable (Samuelson 1983).

8.3 Parasite Virulence

I previously described an epidemiological model of parasite transmission. I repeat that model here and show how it can be analyzed with standard life history methods. Once abstracted, the model demonstrates the essential features of the "tragedy of the commons" problem of social behavior, where there is a conflict between individual success relative to neighbors and the overall success of the group. The abstract form leads immediately to a family of related models for interactions between demography and kin selection.

The standard epidemiological model, used in Section 7.4, is

$$\Delta U = \theta - \delta U - \beta(y) US$$
$$\Delta S = S(\beta(y) U - \delta - z - c),$$

where U and S are, respectively, the number of uninfected hosts available and the number of sick hosts that cannot be infected again. The total population is maintained by new, uninfected hosts, which are recruited at a rate θ. For each contact between uninfected and sick hosts, an individual parasite transmits a progeny at a rate $\beta(y)$, where y is the transmission character of the individual parasite. Parasites die in a sick host when the host dies or the infection is cleared. The clearance rate is c. The natural host death rate is δ. The parasites increase this death rate by the virulence level, z, where z is the average value of y within the host.

The parasite demography can be described in standard life history form as

$$\mathbf{A} = \begin{bmatrix} 0 & \beta(y)/D \\ 1-t & 1-\delta-z \end{bmatrix}, \tag{8.24}$$

where class 1 members are newborn infections and class 2 members are productive parasites in an infected host. The lower-right term is the survival of the parasite to the next time period, where I have taken the clearance rate, $c = 0$. The upper-right term is the fecundity of the parasite, which is given by the transmission rate per new available host (habitat), $\beta(y)$, divided by the term D. The denominator is the effect of density-dependent competition on fecundity (the number of newborns). In the parasite model, density dependence is tied to the number of available hosts, U. As the infection spreads, U declines, and there are fewer open habitats for colonization. The final term, $1 - t$, is the survivorship of newborn infections to the productive stage. This is the latent period of infection. This effect does not occur in the epidemiological model I used, but it is a common part of more complex epidemiological models.

The standard way to handle density dependence is to assume that at equilibrium, $\lambda = 1$. Then, in the normal matrix $\mathbf{A}^*$, with $y = z = z^*$, it is easy to show that

$$D = \frac{(1 - t)\,\beta\,(z^*)}{\delta + z^*}. \tag{8.25}$$

Thus the normal matrix is

$$\mathbf{A}^* = \begin{bmatrix} 0 & (\delta + z^*)\,/\,(1 - t) \\ 1 - t & 1 - \delta - z^* \end{bmatrix}.$$

The normal reproductive values have proportions

$$\mathbf{v} \propto [\,1 - t \quad 1\,],$$

and the frequencies have proportions

$$\mathbf{u} \propto \begin{bmatrix} (\delta + z^*)\,/\,(1 - t) \\ 1 \end{bmatrix}.$$

The derivative of $\mathbf{A}$, evaluated at $y = z = z^*$, shows how components of fitness change

$$\frac{d\mathbf{A}}{dg'} = \begin{bmatrix} 0 & \beta'\,(z^*)\,/\,D \\ 0 & -\tilde{r} \end{bmatrix},$$

where $\beta' = \partial\beta/\partial y$ and, as usual, $\tilde{r} = dz/dg'$. Solving

$$\mathbf{v}\frac{d\mathbf{A}}{dg'}\mathbf{u} = 0$$

yields

$$(1 - t)\,\beta'\,(z^*)\,/D - \tilde{r} = 0,$$

which, from D in Eq. (8.25), is equivalent to (Frank 1992)

$$(\delta + z^*)\,\beta'\,(z^*) - \tilde{r}\beta\,(z^*) = 0. \tag{8.26}$$

This matches the condition presented earlier in Eq. (7.14). A common assumption is to let $\beta(y) = y^s$, with $s < 1$, causing diminishing fecundity returns with increasing virulence. This allows solution of the equilibrium value of the parasite trait

$$z^* = \frac{s\delta}{\tilde{r} - s} \qquad \tilde{r} > s. \qquad (8.27)$$

The same solution can be obtained by analyzing cycle fitness, R_0, in a stable population, $\lambda = 1$, with equilibrium demography. The product of survival to reproductive age and expected adult fecundity can be seen readily from Eq. (8.24) to be

$$w(y,z) = (1 - t)(\beta(y)/D) \sum_{i=0}^{\infty} (1 - \delta - z)^i$$

$$= \frac{(1 - t)\beta(y)}{D(\delta + z)}.$$

Average fitness in a normal population, $w(z^*, z^*)$, is one when D is given by Eq. (8.25). Analysis of $dw/dg' = 0$ yields the same equilibrium as in Eq. (8.26).

The essential features of this model are defined in the **A** matrix in Eq. (8.24). The matrix contains nothing in particular about parasites. Rather, it is a general summary of demography and selection on survival and fecundity components of fitness. The interesting evolutionary tradeoff comes from the fact that an individual's fecundity depends on its own trait, y, whereas the individual's survivorship depends on the average trait value in the local group, z. This tradeoff is also expressed in the cycle fitness, $w(y,z)$.

What if an individual's survival and fecundity depend only on its own trait, y? Analysis of $w(y,y)$ yields the solution in Eq. (8.27) with $\tilde{r} = 1$. The same result obtains when survival and fecundity depend only on average trait values, $w(z,z)$. In both cases there is no tension between individual and group interests. What if fecundity depends on z, and survival on y? Analysis of $w(z,y)$ yields

$$z^* = \frac{s\tilde{r}\delta}{1 - s\tilde{r}} \qquad 1 > s\tilde{r}.$$

In this model, declining relatedness increases individual survival and reduces group fecundity. The opposite pattern occurs in Eq. (8.27), in which declining relatedness reduces group survival and increases individual fecundity. These analyses, taken together, provide a simple family of models that can be applied to different biological interactions (Frank 1996b).

8.4 Social Evolution in Two Habitats

The prior models of survival and fecundity assume that all individuals live in the same kind of habitat. Often, however, some individuals will live in relatively rich habitats, whereas others may find themselves in marginal habitats. Survival and fecundity parameters will, of course, be affected by habitat differences. The different habitats can be thought of as distinct classes, each with its own reproductive values (Holt 1996).

For example, a parasite may attack two kinds of hosts. Suppose, in one host, the infection is highly virulent, and kills the host so quickly that the parasite is rarely transmitted to a new host. The reproductive value of parasites in this host is low; thus selection in this kind of host will have little influence on the parasite's trait values.

Different coefficients of relatedness in the two habitats may interact with demography to determine character evolution. For example, marginal habitats may have low reproductive value and, because individuals are distributed sparsely, relatedness may be high. Crowded, high-quality habitats may have high reproductive value and low relatedness. The net effect is that a certain degree of prudence may be favored by the contribution of lone individuals in marginal areas, balancing the selfish traits favored in the main arena.

This type of problem is simple to formulate with the tools we have developed, but is challenging to analyze completely. The new technical point is how to handle cases in which actors occur in different classes, when there is no conditional adjustment of traits based on class. In this case, the degree of prudence in resource usage is intrinsic to individuals independently of the type of habitat in which they live.

The assumptions are a simple extension of the prior section, with the biology described in the matrix

$$
\mathbf{A}\left(y_q, z_q, y_m, z_m\right) = \begin{bmatrix} 0 & \beta_{qq}\left(y_q\right)/D & 0 & \beta_{qm}\left(y_m\right)/D \\ 1 - t_q & 1 - \delta - z_q & 0 & 0 \\ 0 & \beta_{mq}\left(y_q\right)/D & 0 & \beta_{mm}\left(y_m\right)/D \\ 0 & 0 & 1 - t_m & 1 - \delta - dz_m \end{bmatrix},
$$

where the four classes, given in the columns, are, respectively, juveniles in the high-quality habitat, adults in the high-quality habitat, juveniles in the marginal habitat, and adults in the marginal habitat. The upper-left and lower-right 2×2 blocks show the same demography within habitats as in the one-habitat model in Eq. (8.24). The subscripts on the β's describe where an individual is born and where it lands to begin its life cycle; for example, β_{mq} is a newborn in the high-quality habitat that migrates to the marginal habitat. The parameter $d > 1$ in the lower-right term describes the greater survival penalty that accrues to individuals in marginal habitats. This may be extreme virulence in a secondary host, or a greater reduction in survival when using resources for reproduction.

The normal matrix is $\mathbf{A}^* = \mathbf{A}(z_q^*, z_q^*, z_m^*, z_m^*)$. I take up below whether the character has the same value in the two habitats, that is, whether $z_q^* = z_m^*$.

There are a few conditions that all solutions must satisfy. First, we may take D as a parameter and solve for the population growth, λ, by finding the largest value of λ that satisfies

$$|\mathbf{A}^* - \lambda \mathbf{I}| = 0,$$

where this equation is the standard definition of eigenvalues. Alternatively, we can assume a stable population size, set $\lambda = 1$, and solve for D as in the previous section. Next, we have the class frequencies and reproductive values

$$\lambda \mathbf{u} = \mathbf{A}^* \mathbf{u}$$

$$\lambda \mathbf{v} = \mathbf{v} \mathbf{A}^*.$$

This model can be analyzed in terms of either conditional or unconditional behavior.

CONDITIONAL BEHAVIOR: DIFFERENT TRAITS IN DIFFERENT HABITATS

I provide the steps toward solution, but I do not solve the problem explicitly. The fitness function is

$$w\left(y_q, z_q, y_m, z_m\right) = \mathbf{v} \mathbf{A} \mathbf{u}.$$

We need a simultaneous optimum at $y_q = z_q = z_q^*$ and $y_m = z_m = z_m^*$, for the two independent traits, z_q^* and z_m^*. This bivariate optimum can be studied by the gradient

$$\frac{dw}{dg} = \mathbf{v}\frac{d\mathbf{A}}{dg}\mathbf{u}$$

$$\frac{dw}{dh} = \mathbf{v}\frac{d\mathbf{A}}{dh}\mathbf{u},$$

along with the constraints above. The two traits are controlled by separate predictors; that is, g is the breeding value for the trait expressed in high-quality habitats, and h is the breeding value for the trait expressed in marginal habitats. I have assumed that individual and transmitted breeding values are equal, $g = g'$ and $h = h'$. A local optimum, if it exists, occurs where both derivatives are zero.

In this case, we have several coefficients of correlation. Two terms correspond to standard relatedness coefficients. The value of $\tilde{r}_q = dz_q/dg_q$ is the slope of the average phenotype expressed in the high-quality habitat, z_q, on the breeding value for this trait in the high-quality habitat, g_q. The coefficient $\tilde{r}_m = dz_m/dh_m$ is the slope of the average phenotype expressed in the marginal habitat, z_m, on the breeding value for this trait in the marginal habitat, h_m. By convention, $dy_q/dg_q = 1$ and $dy_m/dh_m = 1$.

Four coefficients of association between characters may also be involved. The coefficients dy_m/dg_q and dy_q/dh_m are the associations between different traits within individuals, typically caused by pleiotropy or linkage disequilibrium. For example, the coefficient $d_{mq} = dy_m/dg_q$ is the effect, within an individual, that the breeding value for the character in the high-quality habitat has on the expression of the character in the marginal habitat.

The other two coefficients are dz_m/dg_q and dz_q/dh_m. These are cross-associations. For example, dz_m/dg_q is the association between the breeding value for the character in the high-quality habitat, g_q, and the average trait value expressed by neighbors within the marginal habitat, z_m. These cross-associations are often the product of associations between neighbors for a single character, and associations between characters within individuals, as $dz_m/dg_q = \tilde{r}_m d_{mq}$.

UNCONDITIONAL BEHAVIOR: SAME TRAIT IN DIFFERENT HABITATS

In many cases, an organism may not adjust its behavior conditionally on habitat type. The "behavior" may be a biochemical trait of a simple organism, such as a virus. An insect or vertebrate may fail to perceive habitat differences, or may lack the machinery to adjust some traits in response to limited information.

Here, $y = y_q = y_m$ is the same trait, controlled by the same breeding value, g. Thus we use the same $\mathbf{A}$ matrix as before, but with $y = y_q = y_m$ and $z = z_q = z_m$. The required derivative is

$$\frac{dw}{dg} = \mathbf{v}\frac{d\mathbf{A}}{dg}\mathbf{u},$$

evaluated at $y = z = z^*$. Expanding the derivative, we follow the usual procedure of labeling breeding value by class of recipient. The terms corresponding to kin selection coefficients are $\tilde{r}_q = dz/dg_q$ and $\tilde{r}_m = dz/dg_m$. It is useful to write the derivative explicitly, using the following labels for classes: the columns of $\mathbf{A}$, as $1, 2, 3, 4$, are, respectively, y, q, μ, m, for juveniles and adults of the high-quality habitat, and juveniles and adults of the marginal habitat. Then

$$\frac{dw}{dg} = (u_q/D)\left(v_y\beta'_{qq} + v_\mu\beta'_{mq}\right)$$
$$+ (u_m/D)\left(v_y\beta'_{qm} + v_\mu\beta'_{mm}\right) \tag{8.28}$$
$$- (c_q\tilde{r}_q + dc_m\tilde{r}_m).$$

The β' terms are the rate of increase in fecundity with increasing trait value, and the terms with relatedness coefficients are the properly scaled reductions in group survival. The standard demographic terms are u, for class frequency; v, for individual reproductive value; and $c = uv$ for class reproductive value. I comment briefly on three aspects of this derivative.

First, a decrease in the trait is prudent in the following sense. Lower trait values reduce the rate of fecundity and increase the rate of survival. If fecundity increases at a diminishing rate with the trait, $\beta'' < 0$, then individual and group fitness are increased by lower trait values. Put another way, the most efficient use of resources is achieved by lowered fecundity and longer survival. This explains why group efficiency rises and equilibrium trait value declines with increasing $\tilde{r}$ in Eq. (8.27). The same trend occurs here in Eq. (8.28), because the derivative is diminished by rising $\tilde{r}$, which implies weaker selection to increase the trait value.

Second, a marginal habitat is, by definition, less productive. If the habitats are equally abundant, then $c_m < c_q$. Thus trait evolution will typically be dominated by selection in the quality habitat (Holt 1996). However, if relatedness is low in the crowded habitat relative to the sparse habitat, $\tilde{r}_q < \tilde{r}_m$, then the marginal habitat can be the dominant force in favoring a degree of prudence.

Third, the relatedness coefficients may be treated as variables or parameters depending on the application. The $\tilde{r}$ coefficients are extrinsic parameters if one has independent data on their values, or if the pattern by which individuals settle into the different habitats and interact with neighbors is independent of survivorship and its consequences. Alternatively, the $\tilde{r}$'s may change with survivorship and demography, as with the density-dependent variable, D. In this case, some constraints must be established to complete the analysis. For example, $\tilde{r}$ in each habitat could be inversely related to numerical abundance in that habitat, or one could use an explicit scheme of migration, settling, and interaction in each habitat.

8.5 Review of the Three Measures of Value

This ends my outline of economic tools required for the study of social evolution. The approach is classical comparative statics. Find a measure of value that is optimized by the processes of interest, and use the powerful mathematical tools of optimization. The currency for natural selection is contribution to the future of the population. Three exchange mechanisms are required.

First, social partners may have correlated trait values. The relation between trait value and contribution to the future population is influenced by such correlations. Maximization of individual reproduction is recovered when one uses the derivative (slope) of partner trait value on the focal individual's predictor value. This direct fitness slope measures consequences of social partners on an individual's reproduction.

Other interpretations of correlation are often used in the biological literature, such as common genealogy or inclusive fitness coefficients. But only a general measure of individual (direct) reproduction influenced by correlated partners can cover all known phenomena without paradox. I described an interesting relationship between slopes that arise naturally in differentiation and the statistical measures of prediction and cause that arise in regression and path analysis (Chapter 4). Maximization by use of differentials turns out to be a limiting case near equilibrium of a general, statistical analysis that follows from the Price Equation.

The second exchange is reproductive value. This measures the expected contribution of each individual to the future of the population. It is often convenient to measure an individual's components of fitness separately. For example, survival and fecundity often require separate reproductive value weightings. From an individual's point of view, survival must be weighted by the individual's reproductive value in the next time period. Current fecundity must be weighted by the reproductive value of offspring in the next time period. Future fecundity must be weighted by the reproductive value of future offspring, multiplied by the probability of survival to produce those offspring.

The final exchange system is marginal valuation. This provides a common currency to measure changes in reproductive value, cost to self, and benefit to partner. Thus the evolution of a social character is not influenced by the reproductive value of the actor and partner, but by the marginal changes in each. Marginal changes require continuity; similar principles can be used for discrete characters (see *Large Deviations*, p. 84).

I apply these economic principles in the following chapters. The topic is sex allocation, the division of resources between sons and daughters. I chose this topic because it illustrates most clearly the simple, economic nature of many problems in social evolution.

9   Sex Allocation: Marginal Value

> I formerly thought that when a tendency to produce the two
> sexes in equal numbers was advantageous to the species, it
> would follow from natural selection, but I now see that the
> whole problem is so intricate that it is safer to leave its solu-
> tion for the future.
> —Charles Darwin, *The Descent of Man*

Sex allocation is the division of resources between male and female. In many animals the problem is how a parent divides its resources between sons and daughters. A hermaphrodite must split resources between the production and transmission of sperm versus the production and nurturing of eggs. Similarly, plants allocate resources separately to pollen and seed.

Consider the structure of the problem from a mother's point of view. She may increase her allocation to sons only by decreasing her allocation to daughters or to future offspring. The gain is measured by the marginal increase in reproduction by sons. This increase depends on how those sons succeed in competition with other males to become the fathers of the next generation. Competition may be against related or unrelated males, requiring correction for value by kin selection. The cost of male investment is measured by the marginal decrease in the success of daughters or of future offspring. These values must be scaled to account for reproductive value among the different classes.

This allocation problem has played a central role in the theory of social evolution. If measures of value work, then one should be able to predict how sex allocation shifts toward male or female with changing demographic and social conditions. The opportunity for comparison is ideal, because predictions take simple forms: more males, larger females, fewer seed-producing flowers, lower production of sperm in some hermaphrodites. These quantities of sex allocation are relatively easy to measure.

The crucial parameters include social interactions among kin, the marginal values of return for male and female investment, and the distribution of limiting resources among individuals. The allocation strategy of each individual may be fixed by genotype, or adjusted conditionally in response to information about resources or relatedness coefficients among social partners. In social groups, there can be conflict among members over how shared resources are allocated to the production of males and females.

Sex allocation has been central to understanding economic principles of social evolution, and to the limits of this theory when applied to real organisms. Many critical summaries of theory and application have been published. Here I emphasize concepts and methods of analysis, with only brief comments on previous theory and application. Good starting points for the literature are Charnov (1982), Frank (1990a), Wrensch and Ebbert (1993), Bourke and Franks (1995), and Crozier and Pamilo (1996).

9.1 Fisher's Theory of Equal Allocation

> If we consider the aggregate of an entire generation of ... off-spring it is clear that the total reproductive value of the males in this group is exactly equal to the total value of all the females, because each sex must supply half the ancestry of all future generations of the species. From this it follows that the sex ratio will so adjust itself, under the influence of Natural Selection, that the total parental expenditure incurred in respect of children of each sex, shall be equal.
> —R. A. Fisher, *The Genetical Theory of Natural Selection*

Fisher's theory of equal investment in the sexes is one of the most widely cited in evolutionary biology. In spite of its popularity, the theory turns out not to apply in the way generally believed. I have discussed this theory extensively in Frank (1990a). In this section I show the standard derivation of Fisher's theory. I then develop in the next section a proper economic framework, in which one can easily see the general features and the limitations of Fisher's idea. I formulate the problem with respect to a mother's division of resources between sons and daughters, but the same conclusions apply to the other types of problems listed above. I use *sex allocation* to refer to the division of resources between males and females. The number of males and females produced is the *sex ratio*.

Shaw and Mohler (1953) developed an expression for the fitness of a mother as a function of her investment in sons and daughters

$$w(y) = \frac{y}{Ny^*} + \frac{1-y}{N(1-y^*)},$$

where y is the fraction of a mother's resources invested in sons, $1 - y$ is the fraction invested in daughters, and y^* is the normal value of the trait in the population.

The number of mothers in the population is N. Thus the fraction of the grandprogeny generation derived from a mother through her sons is given by the first term on the right, and the fraction through daughters is given by the second term. If we solve $dw/dy = 0$, then the equilibrium occurs when $y^* = 1 - y^*$, or $y^* = 1/2$. Thus selection favors equal allocation of resources to the two sexes.

9.2 The Three Measures of Value

The Shaw–Mohler equation contains implicit assumptions about reproductive value, marginal value and kin selection. It is useful to recast the problem to make explicit these three aspects of value.

In the simplest formulation, phenotypes are controlled by the mother, and the recipients are the male and female components of the mother's fitness. The equation for average recipient fitness is

$$W(y) = c_m W_m + c_f W_f, \tag{9.1}$$

where c_m and c_f are the class reproductive values for males and females. The W_m and W_f terms are the recipient fitnesses for a mother's male and female components of reproductive success. These terms are standardized to the same value in a normal population with phenotype $y = y^*$. Thus a change in behavior that causes a change in recipient fitness, W_i, can be interpreted as causing a change in the proportion of the total reproductive value of class i attained by the particular recipients of the behavior.

Following along with our standard method, consider variations in the transmitted breeding value, g', of the trait y. The change in fitness with respect to g' is

$$\frac{dW}{dg'} = c_m \frac{\partial W_m}{\partial y} \frac{dy}{dg'_m} + c_f \frac{\partial W_f}{\partial y} \frac{dy}{dg'_f}. \tag{9.2}$$

We must, as usual, specify the proper interpretation for g' (see *Transmitted Breeding Value*, p. 77). In this case, marginal changes in the mother's success through sex-i offspring, $\partial W_i / \partial y$, are the same as the mother's mates' marginal changes in their corresponding fitness components. Thus we can treat the mother as transmitting both her own genes and her mates' genes to each offspring. This is conveniently handled by defining g'_i as the breeding value for y in sex-i offspring, because offspring are composites of mother's and mother's mates' transmitted breeding value.

The slopes of actor phenotype on recipient's transmitted breeding value are $\tilde{r}_m = dy/dg'_m$ for the association between mother's phenotype and sons' breeding value, and $\tilde{r}_f = dy/dg'_f$ for the association between mothers and daughters. The $\tilde{r}$ terms are direct fitness coefficients. It is traditional to use the inclusive fitness coefficients, which regress recipient (offspring) breeding value on actor (mother) breeding value. I use the inclusive fitness coefficients in the following analysis, although, as I discussed in *Comparison of Direct and Inclusive Fitness*, p. 68, the direct fitness coefficients are more general.

The above substitutions lead to the equilibrium condition $dW/dg' = 0$ as

$$c_m r_m \frac{\partial W_m}{\partial y} = -c_f r_f \frac{\partial W_f}{\partial y},$$

where the derivatives are evaluated at the normal values $y = y^*$. A neater expression can be obtained by first defining $W'_m(y^*)$ as $\partial W_m/\partial y$, evaluated, at its equilibrium value, y^*. Similarly, define $W'_f(y^*)$ as the partial of W_f, evaluated at its equilibrium value. This yields

$$c_m r_m W'_m(y^*) = -c_f r_f W'_f(y^*).$$

On the right side, investment in females declines with y^*, the equilibrium investment in males. I prefer to make the relation between male and female investment explicit by defining $\alpha = 1 - y$ as the fraction of resources invested in females. This allows the substitution

$$\frac{\partial W_f(y)}{\partial y} = -\frac{\partial W_f(\alpha)}{\partial \alpha},$$

so that we can use the equality $W'_f(y) = -W'_f(\alpha)$, where the right side is differentiated with respect to α, leading to the final expression

$$c_m r_m W'_m(y^*) = c_f r_f W'_f(1 - y^*). \tag{9.3}$$

REPRODUCTIVE VALUE

Fisher pointed out that, in symmetric inheritance systems, one-half of the genes come from the mother and one-half come from the father. Thus the class reproductive values of males and females are equal, $c_m = c_f$. The class reproductive value can be expressed as a product of the value of each individual in the class multiplied by the number of individuals in the class, $c = vu$, where v is the value per individual and u is the number in the class (see *Reproductive Value of Each Class*, p. 136). Thus $v_m u_m = v_f u_f$, or

$$\frac{v_m}{v_f} = \frac{u_f}{u_m}.$$

The average reproductive value of a male compared with that of a female is equal to the number of females per male. For example, if there are twice as many females as males, then each male must, on average, have twice as many offspring as each female. This relation shows the powerful frequency dependence of selection affecting sex ratios. This frequency dependence provides an advantage to the rarer sex and tends to equalize the numbers of males and females.

The class reproductive values of males and females are not always equal. For example, in many species the females inherit equally from mother and father, and have two sets of alleles (diploidy), whereas the males inherit a single set of alleles from their mother without paternal input (haploidy). This haplodiploid system occurs in many social organisms, such as bees, ants and wasps. The X chromosome of mammals has the same inheritance pattern. In these systems, two-thirds of the future alleles in the population come from females, and one-third come from males. Thus females have twice the class reproductive value of males, $c_f = 2c_m$.

KIN SELECTION COEFFICIENTS

The kin selection coefficients, r_m and r_f, are the second measure of value in Eq. (9.3). A mother values male and female progeny in proportion to her relatedness to each. With symmetric inheritance, $r_m = r_f$ for sons and daughters. Under haplodiploidy, if mother and father have uncorrelated genotypes, then a mother's relatedness to her son is twice her relatedness to her daughter, $r_m = 2r_f$. This can be understood by noting that a change in the mother's breeding value will have twice as much influence on a son's breeding value as on a daughter's breeding value, because a daughter's genotype is diluted by one-half with uncorrelated alleles from the father. When mother and father are uncorrelated, $c_m r_m = c_f r_f$ for haplodiploidy, as for symmetric systems. Correlation between mother and father breaks the equality for haplodiploidy but not for symmetric systems (see Chapter 10).

Marginal Value

The third measure is marginal value. For many problems, $c_m r_m = c_f r_f$, and Eq. (9.3) reduces to

$$W'_m (y^*) = W'_f (1 - y^*). \tag{9.4}$$

The equilibrium occurs when the marginal increase in male investment provides returns that equal those obtained with a marginal increase in female investment. This type of marginal result was first derived by Charnov et al. (1976); see Frank (1990a) for additional references.

The next step is to consider the form of W_m and W_f. These functions can be written as

$$W_m = \frac{\mu (y)}{\mu (y^*)}$$

and

$$W_f = \frac{\phi (\alpha)}{\phi (\alpha^*)} = \frac{\phi (1 - y)}{\phi (1 - y^*)},$$

where $y = 1 - \alpha$. The normal values of these functions are one when evaluated at $y = y^*$, satisfying the requirement that recipient classes have equal values in a normal population. The equilibrium condition from Eq. (9.4) is

$$\frac{\mu' (y^*)}{\mu (y^*)} = \frac{\phi' (1 - y^*)}{\phi (1 - y^*)}, \tag{9.5}$$

where μ is differentiated with respect to y, and ϕ is differentiated with respect to $\alpha = 1 - y$. This condition shows that equilibrium occurs where the marginal changes in market share are equal for changing investment in males and females (see *Phenotypes and Market Share*, p. 26).

The equilibrium condition in Eq. (9.5) has been studied for various assumptions about the male and female return functions, μ and ϕ. Analysis and summary of prior work is in Charnov (1982) and Frank (1987b). Perhaps the simplest forms assume $\mu(z) = az^s$ and $\phi(z) = bz^t$, with $0 < s, t < 1$, and a, b arbitrary positive constants. This yields the male : female investment ratio, $y^* : 1 - y^*$, as $s : t$.

Equal allocation is favored only when the shapes of the return curves are the same for male and female investment. Thus Fisher's argument about equal reproductive value of males and females explains the strong frequency dependence in sex allocation but fails to account for marginal returns when the sexes differ in response to investment.

9.3 Variable Resources and Conditional Adjustment

The prior model assumed that every individual making an allocation decision had the same total amount of resource. The problem changes significantly when individuals vary in their resource level (Trivers and Willard 1973). For example, in some fish species an individual can function as either a male or a female, but not both. Similarly, some plants can, in each year, produce either seeds or pollen, but not both. Finally, a mother can, in some cases, choose to produce a son or daughter depending on the available resources. Many examples are given in Charnov (1982).

In this scenario, each actor has a different resource level, k. Total allocation to males and females is $ky_k + k(1 - y_k) = k$, where y_k is the fraction of resources allocated to males by an individual with resource level k, and $\alpha_k = 1 - y_k$ is the fraction of resources allocated to females. Each resource level, k, defines a distinct class of actor. Each actor has two recipient classes, males derived from class k actors and females derived from class k actors. Thus total recipient fitness is

$$W = \int c_{mk} W_{mk} f(k) \, dk + \int c_{fk} W_{fk} f(k) \, dk$$

where $f(k)$ is the probability distribution function for individuals with resource level k. I will assume that an actor's relatedness to male and female recipients is the same, $r_m = r_f$; thus we can ignore relatedness coefficients when differentiating. The recipient fitnesses are

$$W_{mk} = \frac{\mu(ky_k)}{\mu(ky_k^*)}$$

and

$$W_{fk} = \frac{\phi(k\alpha_k)}{\phi(k\alpha_k^*)} = \frac{\phi[k(1 - y_k)]}{\phi[k(1 - y_k^*)]}.$$

In a normal population, with phenotypes $\{y_k^*\}$, the normalized recipient fitnesses are $W_{mk} = W_{fk} = 1$. The reproductive values in a normal population are

$$c_{mk} = c_m \frac{\mu(ky_k^*)}{E_\mu}$$

$$c_{fk} = c_f \frac{\phi[k(1-y_k^*)]}{E_\phi},$$

where

$$E_\mu = \int \mu(ky_k^*) f(k) \, dk$$

$$E_\phi = \int \phi[k(1-y_k^*)] f(k) \, dk.$$

If we let $c_m = c_f$, then total recipient fitness is in proportion to

$$W = \frac{\int \mu(ky_k) f(k) \, dk}{E_\mu} + \frac{\int \phi[k(1-y_k)] f(k) \, dk}{E_\phi}.$$

The maximization problem can now be stated explicitly. What is the optimal set of values $\{y_k^*\}$ under the assumption that each class has information about its own resource level, and can adjust its own value of y_k^* independently of other classes? It is useful to separate the problem into three cases.

ALL MALE OR ALL FEMALE BY CONSTRAINT

The allocation may be forced to either all male or all female, that is, y_k^* may be zero or one (Charnov 1982). For example, certain parasitoid wasps can produce one male or one female offspring in a single host. Each host is a separate resource item, requiring a separate decision about sex allocation. The distribution of resources, $f(k)$, is the distribution of host sizes encountered by mothers. For convenience, confine k to the interval $[0, 1]$. The solution is simple if the male and female valuations are equal exactly once, at $k = \lambda$ on the interval $[0, 1]$, that is, if $c_{mk}W_{mk} = c_{fk}W_{fk}$ only at $k = \lambda$. If we write $M(z) = c_{mz}W_{mz}$, and $F(z) = c_{fz}W_{fz}$, then the problem can be summarized by

$$
\begin{aligned}
M(z) &= F(z) & z &= \lambda \\
&> F(z) & z &< \lambda \\
&< F(z) & z &> \lambda,
\end{aligned}
$$

where

$$M(z) = \frac{\mu(z)}{\int_0^\lambda \mu(k)f(k)\,dk}$$

and

$$F(z) = \frac{\phi(z)}{\int_\lambda^1 \phi(k)f(k)\,dk}.$$

Thus males are favored in low-resource classes and females are favored in high-resource classes: $y_k^* = 1$ for $k < \lambda$ and $y_k^* = 0$ for $k > \lambda$. The limits and inequalities are easily switched when males are favored in high-resource classes. Frank (1987b) presented convenient functional forms for μ, ϕ, and f, along with numerical analysis to show how allocation patterns change with changing parameters.

Under the assumption that males are made in the lower classes, and each actor produces only one offspring, the proportion of individuals that are male (numerical sex ratio) is $S = \int_0^\lambda f(k)\,dk$. Frank and Swingland (1988) extended a model by Charnov (1982) to show that the sex produced in the lower classes is always numerically more abundant at equilibrium. The proof is simple. The condition $M(\lambda) = F(\lambda)$ implies

$$\phi(\lambda)\int_0^\lambda \mu(k)f(k)\,dk = \mu(\lambda)\int_\lambda^1 \phi(k)f(k)\,dk.$$

If we make the reasonable assumptions that $\mu(\lambda) > \mu(z)$ for $z < \lambda$, and $\phi(\lambda) < \phi(z)$ for $z > \lambda$, then

$$\phi(\lambda)\mu(\lambda)\int_0^\lambda f(k)\,dk > \phi(\lambda)\mu(\lambda)\int_\lambda^1 f(k)\,dk.$$

Since the integral on the left side is the sex ratio, S, and the integral on the right side is $1 - S$, we have $S > 1/2$. Thus the sex developing under relatively poor conditions will be numerically more abundant.

What about the proportion of resources allocated to males and females? Sometimes that question does not make sense, because the variable, k, that determines sex may be temperature, as in some turtles, crocodiles, and other species (Bull 1983). When k can be interpreted as a resource, the proportion of total resources allocated to males is

$$\Lambda = \frac{\int_0^\lambda kf(k)\,dk}{\int_0^1 kf(k)\,dk}.$$

This allocation ratio may be biased toward either sex, even though the numerical sex ratio is always biased toward the sex of the lower classes (Frank 1987b; Frank and Swingland 1988).

Another pattern that appears to be common is a positive association, across species, between the female : male size ratio and the total female : male allocation ratio. This trend has been observed among some ants, bees, and wasps (Boomsma 1989; Helms 1994). I used numerical analysis of the models outlined in this section to show that a positive association between female : male size ratios and female : male allocation ratios is expected to be common (Frank 1995d).

All Male or All Female Favored by Selection

In the previous section, each actor was constrained to invest only in males or only in females. This constraint applies to each independent investment decision. For example, a mother wasp may lay a male on one host and a female on a different host. If each host defines an independent decision, then the mother is constrained to produce either son or daughter for each decision, but may produce any sequence of the sexes over time. I will discuss the factors that determine the independence of investment decisions in Section 9.4.

What if an actor is free to mix male and female investment? In some cases, selection favors each actor to invest only in one sex, and the results match the analysis in the previous section. This typically occurs when the returns are accelerating for at least one sex. Then high-class actors are favored to invest in the sex with returns that accelerate at a faster rate. At some intermediate resource level, λ, returns are equal for the two sexes. Below that point, the lower-class individuals are favored to invest only in the sex not favored for the high-class actors (see examples in Frank 1987b).

Mixed Allocation Favored in Some Classes

Mixed investment by some actors may be favored when returns are not uniformly accelerating. A class with k resources is favored to mix investment between the sexes if, for some value $0 < y_k < 1$, the marginal returns on male and female investment are equal. If a stable internal point for y_k does not exist for a particular class, then that class invests entirely in males or entirely in females. From the previous sections, the

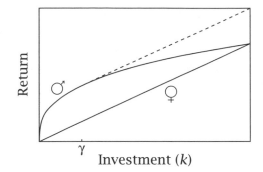

Figure 9.1   Schematic description of all male investment for $k < y$, and constant male investment for $k > y$. Return on male investment increases at a diminishing rate; return on female investment increases at a constant (linear) rate. Up to y, the marginal return (slope) of the male investment curve is greater than the return on female investment. Past y, the marginal return is greater on female investment. The best total return for $k > y$ is shown by the dashed line, which is mixture of y invested in males and $k - y$ invested in females.

fitness of a class with k resources is

$$W_k = \frac{\mu\,(ky_k)}{E_\mu} + \frac{\phi\,[k\,(1 - y_k)]}{E_\phi}.$$

Our usual procedure is to analyze dW_k/dg'_k for each k, where g'_k is the transmitted breeding value of y_k. The condition for mixed investment is obtained from $dW_k/dg'_k = 0$, yielding

$$\frac{\mu'\,(ky_k^*)}{E_\mu} = \frac{\phi'\,[k\,(1 - y_k^*)]}{E_\phi},$$

where μ is differentiated with respect to y_k, and ϕ is differentiated with respect to $\alpha_k = 1 - y_k$. We can write the general condition for mixed investment as $D = C$, where $D = \mu'/\phi'$, the relative rate of return on male investment compared with female investment, and $C = E_\mu/E_f$. If, for a class with k resources, $D > C$ for all values of $0 \le y_k < 1$, then the class is favored to produce only males. Similarly, if $D < C$ for all y_k, then the class is favored to produce only females.

I summarize the case in which there is linear return on female investment and diminishing return on male investment (Fig. 9.1; Frank 1987b). In this case we must solve the simultaneous optimization problem of finding $\{y_k^*\}$ for each class defined by its resource level, k. Linear return on female investment implies $\phi[k(1 - y_k)] = ak(1 - y_k)$, where a is an arbitrary, positive constant. Diminishing return on male investment can be represented by $\mu(ky_k) = b(ky_k)^s$, where $0 < s < 1$ and $b > 0$.

Using these return functions, differentiation yields

$$D = s \, (ky_k^*)^{s-1} .$$

Noting that $s < 1$, the general form of the solution can be described. When k is very small, D is large and greater than C. Thus classes with few resources are favored to produce only males, $y_k^* = 1$. All male investment is favored for $k \leq y < 1$, where y is such that $sy^{s-1} = C$. For $y < k \leq 1$, a mixture of male and female investment is favored. Male investment for each class is $ky_k^* = y$, which implies that $D = C$. Thus total male investment is kept constant with increasing k, and female investment is $k - y$. To summarize, for $0 < k < y$, all male investment is favored. For $y < k < 1$, constant male investment of y and increasing female investment of $k - y$ are favored (Yamaguchi 1985; Frank 1985, 1987b, 1987c). The switch point y decreases as the parameter s declines.

9.4 Returns per Individual Offspring

I have described returns on male and female investment by the functions μ and ϕ, respectively. From a technical point of view, this is sufficient to solve the problem of sex allocation. But how does an actor separate resources into individual packages? For example, a mother must divide resources into individual sons and daughters. Similarly, some plants divide resources into separate male (staminate) and female (pistillate) flowers.

The division of resources into separate offspring is generally known as the size-number problem (Smith 1974). A parent can make fewer large offspring or more small offspring. Sex allocation extends this problem to consider size, number, and sex. The third factor makes the problem more interesting and more difficult, because the choice of sex introduces frequency dependence. The returns for a particular number and size of males depend on the numbers and sizes of males produced by other members of the population. The returns on female offspring depend in a similar way on the size and number of competitors.

The size-number-sex problem can be handled by a two-stage optimization (Frank 1987b). First, take the division between male and female as a parameter fixed extrinsically, and study the size-number problem separately for each sex. Second, under the assumption that each individual follows the optimal size-number split, solve for the optimal division between male and female.

DEFINITION OF INVESTMENT PERIOD

These optimizations depend on the amount of resource available in each investment period. Consider, for example, a parasitoid wasp that lays a single egg on each host that it encounters. If the fitness of offspring depends on the size of its host, then each egg-laying event is a separate investment period for the mother. Whether she makes a son or daughter in one period does not influence her fitness in later periods.

By contrast, the mother may begin the egg-laying sequence with a fixed supply of a crucial resource. The amount she places in each egg may be the primary determinant of offspring fitness. In this case, the entire egg-laying sequence is one investment period.

A proper analysis of investment periods requires study of reproductive value. This allows comparison of the value a mother places on current sons and daughters versus her own survival and future production of offspring. I take this up in Chapter 11. Here I simply assert a certain level of resource for an investment period. I assume the mother is able to divide that resource into any number of sons and daughters, subject to some constraints outlined in the following sections.

FISHERIAN EQUAL ALLOCATION FOR HIGH FECUNDITY PER INVESTMENT PERIOD

The importance of the size-number tradeoff for sex allocation can be shown with a simple example. Suppose that the production of each offspring requires a packaging cost, d, such that investment of d or less provides no return. For example, we can write the return per male offspring as

$$m(\delta_{mi} + d) = \delta_{mi}^s, \tag{9.6}$$

where δ_{mi} is the investment in the ith male produced during the investment period, after the packaging cost d is paid. The parameter $s < 1$ is a parameter that defines diminishing return on investment. Total investment in males during this period is

$$ky = k_m = \sum_{i=1}^{n_m} (\delta_{mi} + d),$$

where n_m is the number of males, k_m is the total male investment, k is the resources available for the investment period, and y is the fraction of those total resources that are provided to males. If we take k_m as fixed, then the problem is to determine the optimal size and number of males that maximizes

$$\mu(ky) = \sum_{i=1}^{n_m} m(\delta_{mi} + d),$$

where one is free to vary both the number of males, n_m, and the investment per male, δ_{mi}. If returns per male diminish with δ_{mi}, as in Eq. (9.6), then it is optimal to invest equally in each male, which implies

$$\delta_{mi} + d = \frac{k_m}{n_m}$$

for all i. The problem is to determine the optimal number of males that maximizes the total return

$$\mu(ky) = n_m \delta_{mi}^s = n_m \left(\frac{k_m}{n_m} - d\right)^s.$$

This maximization is accomplished by solving $d\mu/dn_m = 0$, yielding

$$n_m^* = \frac{k_m(1-s)}{d}. \tag{9.7}$$

The problem with this solution is that the number of offspring must be an integer. For example, if $n_m^* < 1$, then it is best to make one male; if $1 < n_m^* < 2$, then some algebra can determine if one or two is better, and so on. When n_m^* is small, this constraint of discrete packaging has a strong influence on return. As n_m^* becomes large, the difference in total return caused by changing n_m into a nearby integer has vanishing effect. Thus, when k_m/d is large, and consequently n_m^* is large, the total return for total investment in males is

$$\mu(k_m) = \left(\frac{k_m(1-s)}{d}\right)\left(\frac{d}{1-s} - d\right)^s = ak_m. \tag{9.8}$$

In words, when the number of males produced is large, return on male investment, k_m, is linear in k_m, with slope a. Fig. 9.2 shows how number

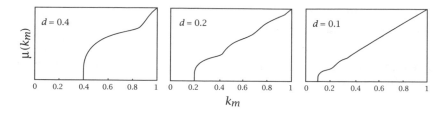

Figure 9.2 Linearization of returns with increasing fecundity per investment period. The maximum of $\mu(k_m)$ was calculated by finding the optimum integer value for number of offspring, n_m^*, which was 2, 4, 7, moving from the left to the right panel. Decreasing d increases the number of offspring, n_m^*, and linearizes the total returns on total investment. The return per offspring was calculated with $s = 0.25$ for all panels.

(fecundity) per investment period influences the shape of the return curve.

The same argument can be applied to females. Suppose, for example, that return per individual female is diminishing, as for males, but with parameter t instead of s. Then, if the number of females per investment period, n_f^*, is large, the total return on total female investment is

$$\phi(ky) = \phi(k_f) = bk_f$$

where b is a constant similar to a in Eq. (9.8). I proved in Section 9.1 that when returns are linear on total male investment and on total female investment, then Fisher's theory of equal allocation follows.

COMPLEXITIES OF LOW FECUNDITY PER INVESTMENT PERIOD

Return scales linearly with number of offspring when fecundity per investment period is high. Linear return leads to Fisher's theory of equal allocation. By contrast, when there is a single offspring per investment period, returns on male and female investment will typically differ. Predicted sex allocation patterns can be derived from the return curves on single male and female offspring and the distribution of resources invested per offspring (see Section 9.3).

Many organisms, such as birds and mammals, have a small number of offspring per investment period. In this case, we have neither the simplification of linear scaling for high fecundity nor the option of using directly the return curves per male and female offspring. The returns per investment period must be constructed from returns per offspring

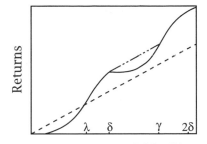

Resources available (k)

Figure 9.3   Returns on investment when a brood contains two males and two females at birth. The sex ratio is constrained, but subsequent investment in each offspring is under parental control. Returns on female investment are assumed to be linear (lower dashed line). Thus separation of resources between the two daughters has no effect on total return. Return for each son follows an S-shaped curve. A parent is favored to use one of five strategies, depending on its resource level, k: for $k < \lambda$, all resources to daughters; for $\lambda < k < \delta$, all resources to one son; for $\delta < k < \gamma$, invest δ in one son and $k - \delta$ in daughters (constant male, shown in dot-dash line); for $\gamma < k < 2\delta$, split resources between sons, with none to daughters; and for $k > 2\delta$, allocate δ to each son and $k - 2\delta$ to daughters. Details in Frank (1987b).

and the distribution of resources available. In addition, the numbers of males and females born in each period is often constrained by the genetic system. A parent has control of investment after birth, but not of the sex ratio itself.

The solution under these complexities is a bit tedious. But we have already used all the required steps. First, solve the size-number problem for each sex, given a fixed amount of resource for that sex. If the birth sex ratio is fixed extrinsically, then the size-number problem is to distribute the limited resources among a fixed number of newborns. The size-number solution yields total return curves for total investment in each sex. Those curves can then be used as in Section 9.3.

Frank (1987b) analyzed an example in detail to illustrate how sex allocation evolves in low fecundity organisms with fixed sex ratio. That example is summarized in Fig. 9.3.

9.5 Critique of the Costs of Males and Females

A common method of empirical study for sex allocation is to compare the "costs" of an individual male versus female offspring and the numbers of males and females (sex ratio). I develop the relevant theory in this section. I show that the comparison by costs is useful when applied to organisms with high fecundity per investment period, but does not make sense for organisms with low fecundity per investment period.

For high fecundity per period, the total male : female allocation is 1 : 1 by Fisher's theory. If the investment per individual offspring is $c : 1$ for male : female, then the ratio of total numbers of males to females in the population (sex ratio) is $1 : c$.

The relation between the cost ratio and sex ratio can be illustrated by the particular example given earlier. The investment per male is obtained by rearranging Eq. (9.7) as

$$\frac{k_m}{n_m^*} = \frac{d}{1-s}.$$

If the same theory is applied to females, with female returns scaled according to the parameter t instead of s, then the investment per female is $d/(1-t)$. Thus we can define c as the ratio of investment in each male relative to the investment in each female

$$c = \frac{1-t}{1-s}.$$

We then have the total male : female population allocation ratio as 1 : 1 by Fisher's theory, the male : female cost ratio as $1 - t : 1 - s$, and the male : female sex ratio as $1 - s : 1 - t$. Equivalently, the cost ratio is $c : 1$, and the sex ratio is $1 : c$. This provides a number of interesting comparative predictions as s, t, and c change.

The situation is different when there is a single offspring per investment period. One sex is often produced under low-resource conditions, whereas the other sex is produced under high-resource conditions (see Section 9.3). Suppose, for example, that males are produced when resources are relatively scarce. Then it may make sense to speak of males as less costly than females. But the investment per male will vary as the resources available vary. Thus there is no single cost value that characterizes males relative to females, but one could use the average cost per sex. I have shown that there is often a positive association between the ratio of average costs and the allocation ratio (Frank 1995d) . For example, as the male : female ratio of average costs increases, the total male : female population allocation ratio increases. This is the opposite of the pattern under Fisherian equal allocation.

Finally, there may be a small number of offspring per investment period. Here the size-number problem is closely associated with the problem of sex allocation. No one has yet shown any clear relation of average investment per male and female to the population allocation ratio or the population sex ratio.

9.6 Multiple Resources

Several different kinds of resources are needed to produce offspring. Only a few theoretical studies have considered the multidimensional nature of sex allocation (reviewed in Rosenheim et al. 1996). Those studies concluded that multiple resources can, in some circumstances, significantly change the pattern and interpretation of sex allocation.

Rosenheim et al. (1996) used solitary bees and wasps as an example of multidimensional sex allocation. When food is abundant, the reproduction of these organisms is often limited by the rate at which they can mature new eggs. Food or nest sites may be the crucial limiting resource at other times and places. Rosenheim et al. (1996) showed that multiple resources combined with temporal and spatial heterogeneity can lead to interesting and testable comparative predictions about shifts in sex allocation. I illustrate this conclusion with an approach that matches the development of the theory in this chapter and differs from the particular models of Rosenheim et al. (1996).

Suppose that a wasp searches for caterpillar hosts. When she finds one, she lays a single egg. That egg develops into an adult by consuming and eventually killing the host. Each host produces a single wasp offspring. Host sizes vary. A daughter gains more in her future fecundity from developing in a large host than a son gains in future mating success. Thus the mothers tend to lay daughters on large hosts and sons on small hosts. Many solitary wasps follow this pattern (Charnov 1982).

When hosts are relatively scarce, a mother's reproductive success is limited by the size and number of hosts she can find. Each egg-laying event is an independent choice because the mother's decision to produce a son or daughter on that host does not affect her future ability to produce additional sons or daughters. The theory of Section 9.3 applies directly. Without particular assumptions, we know that the sex produced on smaller hosts is predicted to be more abundant. From some cases studied numerically, it appears that the total allocation in the population may often be biased toward the sex developing on larger hosts (Frank 1987b, 1995d). Here allocation is measured by host size.

When hosts are very abundant, then a mother's reproduction is limited by the number of eggs she can produce. Because hosts are not limiting, she may use the largest hosts for both sons and daughters. In this case, her return on male investment scales linearly with the number of males she produces. Similarly, she obtains linear return on female investment. Fisher's theory of equal allocation applies.

Suppose hosts are abundant, but a mother has a limited supply of some crucial biochemical product. She starts her reproductive life with a fixed amount of this product, and cannot obtain more during reproduction. If this product limits reproductive success, and lifetime fecundity is high, then she is favored to allocate this resource to sons and daughters according to the theory of costs under Fisherian equal allocation.

10 Sex Allocation: Kin Selection

> The males ... usually complete their life cycle, and die, before
> they are born.
>
> —W. D. Hamilton, "Extraordinary Sex Ratios"

Hamilton's males are mites—small spiderlike organisms. In *Acarophenax tribolii*, mothers carry eggs in their bodies until the eggs hatch and the offspring develop. Sons emerge within the mother's body, mate with sisters, and die. The sisters are then released.

Hamilton (1967) noted two common traits associated with extreme inbreeding in insects and mites. First, most species have very female-biased sex ratios. For example, *A. tribolii* females typically produce just one son and 14 daughters in each brood. Second, many species with inbreeding and biased sex ratios have a haplodiploid genetic system. The sons are asexually produced from the haploid gametes of the mother. A male does not have a father. The diploid daughters are produced by typical mixing of equal haploid gametes from mother and father.

I discuss haplodiploidy in the first section of this chapter. I then consider the sex ratio consequences when related males and females interact in competitive and cooperative ways. The theory explains why biased sex ratios are common with inbreeding species and expands the subject to a wide array of social interactions. I then turn to gamelike interactions between competing mothers who vary the numbers of sons and daughters contributed to a local mating arena. I close by connecting sex ratio and haplodiploidy to the evolution of cooperative societies in insects.

The kin selection aspect of sex ratio has contributed greatly to the concepts and analytical methods of social evolution. The rich array of theoretical predictions is particularly suited to simple experimental and comparative tests. Thus the subject has provided insight into the potential successes and limits of economic analysis in social evolution. Good starting points for the literature include Hamilton (1972, 1979), Trivers and Hare (1976), Charnov (1982), Grafen (1986), Antolin (1993),

Wrensch and Ebbert (1993), Bourke and Franks (1995), and Crozier and Pamilo (1996).

10.1 Haplodiploidy

RELATEDNESS AND REPRODUCTIVE VALUE

Early formulations of sex allocation commonly combined relatedness and reproductive value into a single measure of total value (Hamilton 1972). Taylor (1988b) showed that it is essential to split these components because they measure independent dimensions of value. Taylor also showed that patterns of genetic transmission, such as diploidy and haplodiploidy, can be described by the matrix formulations given in Chapter 8. For example, a diploid system has movement of genes described by

$$\mathbf{A} = \frac{1}{2} \begin{bmatrix} u_f & \alpha u_f \\ u_m & \alpha u_m \end{bmatrix},$$

where contribution of sex-j parents to sex-i offspring is given by a_{ij}. I adopt the convention that females are in row 1 and column 1, and males are row 2 and column 2. The terms u_f and u_m are proportional to the numbers of daughters and sons produced by each mother. The sex ratio, measured as number of females per male, is $\alpha = u_f / u_m$, which is also the average number of mates per male. Thus each male produces, on average, αu_f daughters and αu_m sons. The one-half in front of the matrix accounts for the fact that, in this diploid genetic system, each offspring receives one-half of its genes from its mother and one-half from its father.

I use the methods of Chapter 8 to analyze the properties of this $\mathbf{A}$ matrix. First, note that the growth rate of the population is the number of daughters born to each mother, $\lambda = u_f$, so $\mathbf{Au} = u_f\mathbf{u}$. Thus, to stabilize the size of the population, assume that each mother produces one surviving daughter (divide $\mathbf{A}$ by u_f), yielding

$$\mathbf{A} = \frac{1}{2} \begin{bmatrix} 1 & \alpha \\ 1/\alpha & 1 \end{bmatrix}.$$

We now have $\mathbf{Au} = \lambda\mathbf{u}$, with dominant eigenvalue $\lambda = 1$. We find the individual reproductive values of females and males, v_f and v_m, by solving $\mathbf{vA} = \lambda\mathbf{v}$. The outcome is $v_f u_f = v_m u_m$. Recall that the reproductive value of a class is $c_i = v_i u_i$, the product of the reproductive value per individual, v_i, and the relative frequency of individuals of that class, u_i. Thus, $v_f u_f = v_m u_m$ implies that the total (class) reproductive values of females and males are equal in a diploid genetic system, $c_f = c_m$.

In haplodiploid systems, the diploid daughters are produced by typical mixing of equal haploid gametes from mother and father. The sons are asexually produced from the haploid gametes of the mother. Thus a father gets zero credit for sons. A mother gets twice the credit for a son as in a diploid system, because haplodiploid sons are assigned fully to the mother, rather than splitting credit between mother and father. A stable haplodiploid population is described by

$$\mathbf{A} = \frac{1}{2} \begin{bmatrix} 1 & \alpha \\ 2/\alpha & 0 \end{bmatrix},$$

which yields a ratio of class reproductive values, $c_m : c_f$, as $1 : 2$ (Price 1970). Thus an allele in the distant future has a $1/3$ probability of residing in a male today.

Asymmetric genetics, such as haplodiploidy, often cause different reproductive value weightings for the sexes. Taylor (1988b) emphasized that symmetric genetics with asymmetric parenting can also cause a bias in reproductive value. For example, suppose that diploid males and females reproduce in each generation. The females die, and the males survive to the following generation with probability s. The ratio of surviving to newborn males is $s : 1 - s$. Assume, among newborns, that the ratio of females to males is $1 : 1 - s$. The s surviving males complement the $1 - s$ newborn males, so that there is a stable sex ratio of $\alpha = u_f/u_m = 1$.

A male's average contribution to the females of the following generation is $1/2$. The males' expected contribution to the males of the following generation is $s + (1 - s)/2 = (1 + s)/2$; that is, they contribute directly by surviving to form a fraction s of the next generation, and they contribute one-half of the genes to the fraction of newborn males, $1 - s$. Similarly, a mother's expected contribution to females of the next generation is one-half, and her expected contribution to males is one-half of the fraction of newborns, for a total of $(1 - s)/2$. The parent-offspring matrix is therefore

$$\mathbf{A} = \frac{1}{2} \begin{bmatrix} 1 & 1 \\ 1 - s & 1 + s \end{bmatrix}.$$

The ratio of reproductive values is

$$\frac{c_m}{c_f} = \frac{1}{1-s} = 1 + s + s^2 + \ldots$$

The extra reproductive value of males comes from their direct contribution to future generations by surviving to reproduce at a later time.

This last example demonstrates the need to separate reproductive value weightings from the genetic system, such as diploidy or haplodiploidy. In the remainder of this chapter I present only simple demographics, so that $c_m : c_f$ is 1 : 1 for diploidy and 1 : 2 for haplodiploidy. I analyze complex demography and reproductive value in the following chapter.

Mechanism of Conditional Sex Ratio Adjustment

Haplodiploidy gives the mother the potential to control the sex of each offspring (Bull 1983). If she lays an unfertilized egg, it develops into a haploid son. If she fertilizes an egg, it develops into a diploid daughter. By contrast, diploid systems rarely seem to allow precise control over the sex of each offspring (but see Yamaguchi 1985). Nagelkerke (1993) describes some complexities of haplodiploid genetics and sex ratio control.

The haplodiploid groups of insects and mites often have spatially localized breeding structures in which male and female relatives compete or cooperate (Wrensch and Ebbert 1993). Kin selection models of sex allocation have been developed with considerable success for this group. The next section describes the basic structure of sex allocation theory for competitive and cooperative interactions among relatives. The following section analyzes individual mothers who use information about the correlated strategies of neighbors to adjust their own sex allocation.

10.2 Competitive and Cooperative Interactions among Relatives

The basic equation for recipient fitness is

$$W(y, z) = c_m W_m + c_f W_f$$

which extends Eq. (9.1) by making fitness a function of both the actor's allocation, y, and the average allocation of actors in the local group,

z. The actor's allocation to males is y, and its allocation to females is $1 - y$. The average allocation by actors in the local group is z to males and $1 - z$ to females. Differentiating with respect to small variations in transmitted breeding value, as in Eq. (9.2), yields

$$\frac{dW}{dg'} = c_m \left(\frac{\partial W_m}{\partial y} \frac{dy}{dg'_m} + \frac{\partial W_m}{\partial z} \frac{dz}{dg'_m} \right)$$
$$+ c_f \left(\frac{\partial W_f}{\partial y} \frac{dy}{dg'_f} + \frac{\partial W_f}{\partial z} \frac{dz}{dg'_f} \right).$$

Replacing phenotypic derivatives by direct fitness relatedness coefficients yields

$$\frac{dW}{dg'} = c_m \left(\tilde{r}_m \frac{\partial W_m}{\partial y} + \tilde{R}_m \frac{\partial W_m}{\partial z} \right) + c_f \left(\tilde{r}_f \frac{\partial W_f}{\partial y} + \tilde{R}_f \frac{\partial W_f}{\partial z} \right). \qquad (10.1)$$

I drop the "~" over the relatedness coefficients and use the inclusive fitness coefficients in the following discussion. The inclusive fitness coefficients have the advantage that causal interpretations can be made consistently from the actor's point of view. These coefficients are, however, less general than the direct fitness form (see Section 4.4).

In Eq. (10.1), the terms r_m and r_f are the relatedness coefficients of the actor (mother) to its own components of male and female fitness (sons and daughters). The terms R_m and R_f are the relatedness coefficients of an actor to the sons and daughters of a randomly chosen mother whose fitness is affected by the actor's sex ratio. Note that a mother's fitness components are affected by her own sex ratio, so the actor is included in the potential set of random recipients.

The model is clarified by first deriving the classic Hamilton (1967) result for local mate competition, and then by expanding the explanation for the model given earlier in Eq. (8.9).

The Hamilton model assumes that a mother lands on a patch and makes some sons and daughters. A few mothers lay eggs in each of the many discrete patches in the population. Offspring emerge in each patch, mating occurs locally by competition among the males, and the females do not compete for resources. After mating, the females disperse to find a new patch and start the cycle again. Investment in sons and daughters is measured by numbers of offspring of each sex; thus y is the fraction of a mother's offspring that are male, and z is the fraction of the local group's offspring that are male. A mother's fitness through sons is

$$W_m = y \left(\frac{1-z}{z} \right) = \frac{y}{z} (1-z),$$

where the average number of mates per male in the local group is $(1 - z)/z$, and the number of sons produced by the actor is proportional to y. Alternatively, one can think of y/z as the competitive success of a mother through males, y, compared to the average competitive success of neighboring mothers (including herself), z. Thus, y/z is relative fitness through sons that compete locally for mates, and $1 - z$ is proportional to the total number of mates available. Note the equivalence of fitness through sons, W_m, with models in *Tragedy of the Commons*, p. 130.

A mother's fitness through females is simply proportional to the number of female offspring because there are assumed to be no competitive or cooperative interactions among females, thus

$$W_f = 1 - y.$$

From Eq. (10.1), $dW/dg' = 0$ evaluated at $y = z = z^*$ yields

$$z^* = \frac{c_m (r_m - R_m)}{c_m r_m + c_f r_f}. \tag{10.2}$$

This result is very general because it applies to any genetic system and pattern of migration for which we can calculate the class reproductive values, c_m and c_f, the relatedness of mother (or other actor) to male and female offspring, r_m and r_f, and the relatedness of the actor to males in the local breeding group, R_m (Taylor 1988b).

Hamilton (1967) did not have access to this analytical technology. He was forced to assume, as for diploidy, that $c_m = c_f$ and $r_m = r_f$. He did not analyze the problem in terms of kin selection. Instead, he assumed that in each patch there are N unrelated mothers, so that a mother's relatedness to male offspring in the group is $R_m = r_m/N$; that is, she is related to the fraction $1/N$ of the offspring that are hers, by r_m, and unrelated to the others. These substitutions in Eq. (10.2) yield Hamilton's (1967) classic formula for local mate competition

$$z^* = \frac{N-1}{2N}.$$

Eq. (10.2) is easier to understand when it is expressed, from the actor's point of view, in terms of the value of each unit investment in male and female fitness components (Frank 1986b, 1986c, 1987c). If we define the equilibrium ratio of females and males as $F/M = (1 - z^*)/z^*$, then we can rearrange any equilibrium for z^* and express it as

(F/M) (Marginal value per male) = Marginal value per female,

where male value is adjusted by F/M because the breeding success of each male depends on this ratio. Thus the equilibrium sex ratio $M : F$ can be derived and expressed as

Marginal value per male : Marginal value per female.

For example, Eq. (10.2) can be rearranged into this ratio form as

$$c_m r_m - c_m R_m : c_f r_f + c_m R_m. \tag{10.3}$$

An actor obtains marginal value for producing an extra male as follows. The term $c_m r_m$ is the direct value of that male. This direct value must be diminished by the cost the extra male imposes on the mating success of neighboring males, $c_m R_m$, when competing for the fixed number of females in the local group. All marginal benefits and costs are scaled for reproductive value and relatedness. The marginal value of an extra female can be read similarly. The direct value of the female is $c_f r_f$. In addition, the extra female increases the average mating success of the males in the local group, providing a benefit to those males in proportion to $c_m R_m$ when analyzed from the actor's point of view.

The valuation form shows that the model combines mating competition and mating benefits between male and female relatives. This is easy to see by comparing with a model in which females disperse before mating (Frank 1986c). In this case male fitness depends on the number of mates available, determined by the population average $1 - z^*$, rather than the local number $1 - z$, so that

$$W_m = \frac{y}{z}(1 - z^*),$$

and the equilibrium valuation is

$$c_m r_m - c_m R_m : c_f r_f.$$

This is a result for pure local mate competition.

It may also be that female relatives compete for limited resources, for example

$$W_f = \frac{1-y}{1-z}.$$

For males, I use the prior model in which mating is local, and the number of mates is proportional to $1 - z$. Because we have changed the scaling for W_f, we must also rewrite W_m as

$$W_m = \left(\frac{y}{z}\right)\left(\frac{1-z}{1-z}\right) = \frac{y}{z}.$$

This yields the relative valuation

$$c_m r_m - c_m R_m : c_f r_f - c_f R_f.$$

The effect of female competition among relatives, $-c_f R_f$, is referred to as local resource competition (Clark 1978). With this form of local resource competition, the value that an extra female provides to male relatives has disappeared because each additional female mate is matched by a decline in productivity per female.

The combined effect of local mate competition, local resource competition, and mating bonus can be seen from the earlier example in Eq. (8.9). The life cycle began with mating on the local patch, followed by partial dispersal of mated females and competition among females for breeding sites. The male and female fitnesses were

$$W_f(y,z) = (1-y)\left[(1-\mu)\,p\,(z) + \mu\,(1-c)\,p\,(z^*)\right],$$

where

$$p(z) = \left[\frac{1}{(1-z)(1-\mu) + (1-z^*)\mu(1-c)}\right]$$

is the breeding probability of a female who competes on a z-patch, with μ as the dispersal rate and c as the cost of dispersal. Similarly, the fitness of a male offspring is

$$W_m(y,z) = \frac{y}{z}(1-z)\left[(1-\mu)\,p\,(z) + \mu\,(1-c)\,p\,(z^*)\right].$$

The ESS in ratio form is

$$c_m r_m - c_m R_m : c_f r_f + c_m R_m - k^2\,(c_f R_f + c_m R_m),$$

where $k = (1 - \mu)/(1 - c\mu)$ is the probability that a mated female is native to her breeding patch. We have the same terms as in previous models, but now the effect of local resource competition is $-k^2 c_f R_f$. The term k^2 is the probability that an actor's daughter and a random female offspring remain in the patch and compete, and R_f is the actor's relatedness to a random female offspring in the patch before dispersal. A similar explanation applies to the mating bonus term on the female value side, $c_m R_m$. An extra female provides a mating bonus of $c_m R_m$ during the mating phase, but reduces the mating bonus provided to a related male by k^2, the probability that related males' gametes and the extra female are competing in the same patch after dispersal.

All of these solutions are given with relatedness coefficients as parameters. The underlying system of dispersal and genetics can be used to calculate the relatedness coefficients, as explained in Section 7.2. The calculations are tedious but simple for complex migration schemes and asymmetric genetic systems, such as haplodiploidy. Several examples are given in Frank (1986b, 1987c) and Taylor (1988b).

10.3 Sex Ratio Games

Suppose two mothers lay their eggs in an isolated patch, with local mating among offspring. The number of grandchildren produced by each female depends on her own sex ratio and the sex ratio and relative brood size of her partner. Hamilton (1967) noted the explicit, two-player-game structure of this interaction. The situation becomes particularly interesting when the paired females have different information about each other and can adjust their sex ratio in response to that information.

Sex ratio interactions between isolated females often arise in certain kinds of parasitic wasps. A female finds a host and lays some male and female eggs. A second female comes along and is able to detect that the host has already been parasitized. This second female can then adjust her sex ratio given the information that another brood is already present.

This sex ratio problem is important because predictions can be tested by experimental and comparative study. Once again, sex ratio provides a touchstone for understanding the processes that influence social behavior, with unique opportunities to test the limits of applicability in nature. Variations of these sex ratios games have been analyzed by Hamilton

(1967), Suzuki and Iwasa (1980), Werren (1980, 1983), Charnov (1982), Frank (1985), Herre (1985), Stubblefield and Seger (1990), and Nagelkerke (1993).

I begin with two simple extensions of the standard local mate competition model that led to Eq. (10.2). First, brood sizes vary, but each female is unable to assess her own relative contribution to her group. Second, each female is able to assess her relative contribution to the local group and adjust her sex ratio in response to this information.

In the first, unconditional model, we can separate the actors (mothers) into distinct classes based on their relative contribution to the group. Define $\beta_i = k_i/(n\bar{k})$ as the relative contribution of the ith female in a group with n females, where k_i is the brood size of the ith female and $n\bar{k}$ is the total brood size of all females. In this model, each female in an equilibrium population produces the same sex ratio. Thus an i-type female's contribution to future generations is directly proportional to β_i. The fitness that must be maximized is therefore proportional to

$$W = \sum_{i=1}^{n} \beta_i \left(w|\beta_i\right), \qquad (10.4)$$

where $w|\beta_i$ is the fitness of a female given that her relative contribution is β_i. This fitness is

$$w|\beta_i = c_m \left(W_m|\beta_i\right) + c_f \left(W_f|\beta_i\right).$$

Let the actor's sex ratio be y for frequency of males, and let the average sex ratio of all other females, excluding the actor, be z. Use similar definitions subscripted by f when the recipient class is female. Then

$$W_m|\beta_i = \left(\frac{y}{\beta_i y + (1 - \beta_i)\, z}\right) (1 - \beta_i y - (1 - \beta_i)\, z)$$

and

$$W_f|\beta_i = 1 - y.$$

All recipient fitnesses are $1 - y$ in a normal population in which $y = z$. Differentiating Eq. (10.4) with respect to small deviantions in transmitted breeding value, setting $dW/dg' = 0$, and solving at $y = z = z^*$ yields the ratio $z^* : 1 - z^*$ as

$$c_m r_m - c_m R_m : c_f r_f + c_m R_m$$

which is identical to the standard local mate competition result given previously in Eq. (10.3). The only difference is that the relatedness of an actor to a random male offspring on the patch is

$$R_m = \frac{r_m + (n' - 1) R'_m}{n'} \qquad (10.5)$$

where R'_m is the average relatedness of an actor to a male offspring that is not her own, and n' is an effective size of the local group defined as (Frank 1985)

$$1/n' = \sum_{i=1}^{n} \beta_i^2.$$

If mothers are unrelated to the sons of the other $n - 1$ females who lay eggs in the same patch, then $R'_m = 0$, and we can write the equilibrium frequency of males as

$$z^* = \frac{c_m r_m}{c_m r_m + c_f r_f} \left(\frac{n' - 1}{n'} \right).$$

The specific calculations for r_m and r_f under haplodiploidy and this mating system are given in Frank (1985).

In the second case, each female has knowledge of her own β_i and can adjust her sex ratio conditionally on this information. Here we optimize simultaneously $w|\beta_i$ by solving $d(w|\beta_i)/dg'_i = 0$ for $i = 1, \ldots, n$. If mothers are unrelated to the sons of the other $n - 1$ females who lay eggs in the same patch, then

$$n\beta_i z_i^* = z^* = \frac{c_m r_m}{c_m r_m + c_f r_f} \left(\frac{n - 1}{n} \right),$$

where z_i^* is the sex ratio of the ith mother with relative contribution β_i, and z^* is the average sex ratio of all mothers. This result requires that $\beta_i > z^*$ for all i. Derivations can be found in Frank (1985) and Yamaguchi (1985), with extensions by Stubblefield and Seger (1990).

The interesting point in this conditional model is that each female produces the same number of males independently of her brood size. Constant male behavior occurs because local mate competition causes diminishing returns on male investment, whereas the lack of resource competition among females leads to linear returns on female investment. This pattern was illustrated in Fig. 9.1. If there were local resource competition among females and no competition among related males, then the equilibrium would be a constant number of females produced independently of brood size (Frank 1987c). The reason is the same in the constant female case: diminishing returns on female investment and linear returns on male investment.

SEQUENTIAL GAME

Suppose a wasp parasitizes a host, and a second wasp parasitizes the same host at a later time with probability p (Hamilton 1967). I discuss two cases. The first assumes that neither wasp is able to assess whether a host has been parasitized previously. The second case assumes each female is able to detect whether it is first or second and adjust its sex ratio based on this information.

All mating occurs on the host, so there is local mate competition among males. The mated females fly off to find new hosts. I assume that there is no resource competition among females during development.

I begin with the unconditional case. The sum of recipient fitnesses is taken over the different classes. Here, one type of class division is whether a female is alone or together with another female. The weights are the probability that a future allele comes from a class in the present, that is, the class reproductive values. Thus the total recipient fitness can be written as

$$W = (1 - q) W_a + qW_t,$$

where W_a is a female that is alone in a host, and W_t is a female together in a host with another female. There are $1 - p$ and p hosts with females alone or paired, respectively, and two females in each paired host, so the relative weights are $1 - q = (1 - p)/(1 + p)$ and $q = 2p/(1 + p)$. In this unconditional model the fitness of females in the normal population is the same in the two types of hosts.

The recipient fitness for a lone female is

$$W_a = c_m W_{am} + c_f W_{af}$$
$$= c_m (1 - y) + c_f (1 - y)$$

because male fitness depends on the number of available mates, $1 - y$, and daughter fitness is the number of daughters produced, $1 - y$. When a female is together with another female, the recipient fitness is

$$W_t = c_m W_{tm} + c_f W_{tf}$$
$$= c_m \left(\frac{y}{y + z} \right) (1 - y + 1 - z) + c_f (1 - y),$$

where the y's are the actor's phenotypes and the z's are the partner's phenotypes. The equilibrium is obtained by solving $dW/dg' = 0$ at $y = z = z^*$. I drop the "~" on the relatedness coefficients and use inclusive fitness coefficients, yielding

$$z^* = \frac{qc_m (r_m - R'_m)}{2 (c_m r_m + c_f r_f)}$$

where R'_m is the relatedness of a mother to the sons of its partner when paired with a partner in a host. We can replace R'_m by R_m, the relatedness of a female to a randomly chosen male offspring when paired with a partner. From Eq. (10.5) with $n' = 2$, the replacement is $R'_m = 2R_m - r_m$. Using this substitution and writing the result as the male : female ratio, $z^* : 1 - z^*$, yields

$$qc_m (r_m - R_m) : c_f r_f + (1 - q) c_m r_m + qc_m R_m.$$

This result can be read as follows. On the left, the marginal value of an extra son when a mother is paired, with probability q, is the standard local mate competition valuation of $c_m(r_m - R_m)$. When the mother is alone, with probability $1 - q$, additional sons provide no value because they only compete with sons already present. On the right, we have the standard value of a daughter, $c_f r_f$, plus the value an extra daughter provides to her mates, $c_m r_m$, when the mother is alone, with probability $1 - q$, and the value an extra daughter provides to a random male in the group, $c_m R_m$, when the mother is paired, with probability q. The equilibrium result ignores the complications that arise when the expected number of surviving sons is less than one (for an overview, see Nagelkerke 1993).

In the second model, the females can assess whether another female has already laid eggs on a particular host. I study two traits, which I assume to be uncorrelated within each individual. The first trait is the sex ratio produced in a host given that it has not been previously parasitized. The second trait is the sex ratio produced in a previously parasitized host. The method is to solve for a simultaneous equilibrium for the two traits.

For the first female who finds an unparasitized host, total fitness is

$$W_1 = c_m W_{1m} + c_f W_{1f}$$

$$= c_m \left[(1 - p) (1 - y) + p \left(\frac{y}{y + z} \right) (1 - y + 1 - z) \right] + c_f (1 - y),$$

where y is the sex ratio of the first female, and z is the sex ratio of the second female who parasitizes the host, with probability p. For the second female total fitness is

$$W_2 = c_m W_{2m} + c_f W_{2f}$$

$$= c_m \left(\frac{z}{y+z} \right) (1 - y + 1 - z) + c_f (1 - z).$$

Each parasitized host has a first female, and p parasitized hosts have a second female. Thus total recipient fitness is

$$W = W_1 + pW_2$$

$$= c_m (W_{1m} + pW_{2m}) + c_f (W_{1f} + pW_{2f})$$

$$= c_m W_m + c_f W_f.$$

It is easy to show from the above definitions that $W_m = W_f$ in a normal population with $y = y^*$ and $z = z^*$. This provides a check on the method.

The solution is obtained by solving for y^* and z^* from $dW_1/dg'_1 = dW_2/dg'_2 = 0$. The breeding value, g_1, affects the sex ratio, y, of the first female to parasitize a host. The breeding value, g_2, affects the sex ratio, z, of the second female to parasitize a host.

The direct fitness coefficients for offspring to their own mother are $\tilde{r}_m = dy/dg'_{1m} = dz/dg'_{2m}$ and $\tilde{r}_f = dy/dg'_{1f} = dz/dg'_{2f}$, where I have assumed that these coefficients are the same for first and second females. The direct fitness coefficient for a female's sons to the sex ratio of her partner is $\tilde{R}'_m = dz/dg'_{1m} = dy/dg'_{2m}$. I assume that the two traits are uncorrelated, thus $\tilde{R}'_m = 0$ even if the paired mothers are related in a genealogical sense. Note that there is no corresponding inclusive fitness coefficient for $\tilde{R}'_m$. For example, this coefficient measures association between breeding value for sex ratio as first female and the phenotype of partners who lay as second females. This association is between two different characters.

Replacing the direct fitness coefficients by their matching inclusive fitness forms, the equilibrium trait value for first females is

$$y^* = \left(\frac{2c_m r_m}{c_m r_m + c_f r_f} \right) \left(\frac{p^2}{(1+p)^2} \right),$$

and the equilibrium trait value for second females is

$$z^* = y^*/p.$$

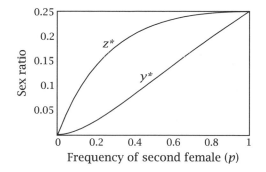

Figure 10.1   Sex ratios of the first female to parasitize a host, y^*, and the second female to parasitize a host, z^*. Females can assess whether a host has been parasitized previously.

This solution was first given by Hamilton (1967). I have extended Hamilton's formulation here to handle the asymmetric class reproductive values and relatedness coefficients, c and r, that arise in haplodiploids with this type of breeding system. The approach here can also handle correlations between partners and the potential for kin recognition by keeping track of the R'_m terms.

If we assume that the organism has a symmetric genetic system, such as diploidy, then $c_m r_m = c_f r_f$, and the solutions for y^* and z^* are given in Fig. 10.1.

SEQUENTIAL GAME WITH VARIABLE BROOD SIZE AND DISPERSAL

The brood size of a second (or later) female influences her favored sex ratio (Werren 1980). For simplicity, let us assume that $c_m r_m = c_f r_f$, so that we can drop the m and f subscripts. Let the sex ratio of the all prior females be fixed at y^*, the sex ratio of the last female to lay on a host be a variable, z, and k be the brood size of the last female on a host relative to all prior females. The last female's fitness is

$$w = \left(\frac{kz}{y^* + kz}\right)[1 - y^* + k(1 - z)] + k(1 - z). \tag{10.6}$$

We obtain the equilibrium sex ratio of the last female, given the sex ratio of the prior females and the relative brood size, k, by solving $dw/dg' = 0$, and assuming, as usual, that the slope of an individual's phenotype

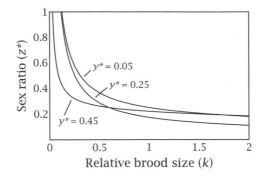

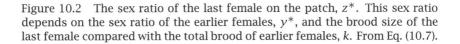

Figure 10.2 The sex ratio of the last female on the patch, z^*. This sex ratio depends on the sex ratio of the earlier females, y^*, and the brood size of the last female compared with the total brood of earlier females, k. From Eq. (10.7).

on its own transmitted breeding value is $dz/dg' = 1$. This yields

$$z^* = \frac{-y^* + \sqrt{y^{*2} + y^*(1 + k - 2y^*)/2}}{k}. \tag{10.7}$$

Suppose the prior females' sex ratio is biased toward daughters, $y^* < 1/2$, because of local mate competition. Then the last female is favored to make mostly sons if she lays a relatively small brood, and to increase her allocation to daughters with increasing brood size (Fig. 10.2; Werren 1980). The first few sons gain the mating advantage provided by the biased sex ratio, each expecting $(1 - y^*)/y^* > 1$ successful matings. With a small brood, the female's sons will compete mostly against sons of the prior females rather than against each other. As the female's number of sons increases, the mating advantage caused by the biased sex ratio declines, and competition among sons partially offsets the mating advantage. Thus she is favored to increase her allocation to daughters.

Nagelkerke (1994) emphasized that for the last female laying on a patch, the fitness per offspring is higher for small brood sizes (Fig. 10.3). This occurs because the early sons gain the mating advantage described above while avoiding local competition against brothers. A female is therefore favored to lay a small number of males in many different patches. Spreading eggs among patches allows sons to avoid competition with their brothers, and is similar to the dispersal model in Section 7.2 (see also Crespi and Taylor 1990). When there is a cost for moving between patches, then a female faces the tradeoff between the

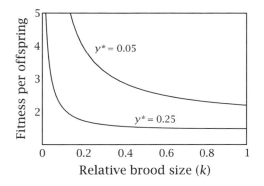

Figure 10.3   Fitness per offspring for the last female on the patch. Fitness was calculated by w/k from Eq. (10.6), using the values of z^* in Eq. (10.7).

costs of search and dispersal and the benefits of small brood size. A simple extension of Eq. (10.6) describes this tension

$$w = n \left[\left(\frac{kz}{y^* + kz} \right) [1 - y^* + k(1 - z)] + k(1 - z) \right],$$

where n is the number of patches visited, and the total brood size over all patches is

$$nk = K - (n - 1)c,$$

with K as the maximum brood size when all eggs are placed in a single patch, and c as the cost of travel between patches. With these definitions, one can maximize w to obtain the simultaneous maximum for sex ratio, y^*, and number of patches to visit, n^*.

10.4 Social Topics

Most social insects have a haplodiploid genetic system. The social haplodiploid species include bees, ants, and wasps (Hymenoptera). Other complex social species have symmetric, diploid inheritance, including termites and naked mole rats. There is an extensive literature on the biology of these social groups (Wilson 1971; Sherman et al. 1991). I briefly summarize two topics concerning sex allocation in the haplodiploid social insects (Crozier and Pamilo 1996). First, there is a conflict between the workers and the queen over the allocation of resources to male and female reproductives. Second, variation in sex allocation among families may have played an important role in the evolutionary origins of complex sociality. I cover aspects of reproductive value and sociality in Chapter 11.

CONFLICT BETWEEN QUEEN AND WORKERS

Queens lay most of the eggs in social insect colonies. A haplodiploid queen can control the sex ratio of offspring by adjusting the frequency of eggs that are fertilized and develop into diploid daughters and the frequency of unfertilized eggs that develop into haploid sons. The workers tend the eggs and raise the offspring. The workers' control over investment in eggs may allow them to adjust subsequent allocation of resources into male and female offspring.

The queen and workers differ over the favored split between male and female investment (Trivers and Hare 1976). In the simplest case, with equivalent marginal returns on male and female investment, and no local mate competition or local resource competition, the favored ratio of male : female investment is

$$c_m r_m : c_f r_f.$$

For haplodiploidy and simple life cycles, $2c_m = c_f$, so the equilibrium allocation ratio is $r_m : 2r_f$. The r coefficients are regressions between actor and recipient. The important point is that the r coefficients differ depending on whether the queen or the workers control sex allocation. The different r coefficients cause a conflict between the queen and workers over the allocation of resources to male and female reproductives.

The queen is the actor and the recipients are son and daughter when the queen controls allocation of resources. Alternatively, the workers are the actor class and the recipients are the queen's offspring when the workers control the allocation to males and females. The workers are typically the queen's daughters, and the subsequent male and female offspring are the workers' brothers and sisters.

The equilibrium sex allocation is calculated from $r_m : 2r_f$. The calculation of relatedness coefficients is simplest when we use inclusive fitness coefficients based on genealogy. The genealogical calculation is given in Fig. 10.4. When the queen is in control, and there is no inbreeding ($f = 0$), then her favored male : female investment ratio is 1 : 1. By contrast, the workers' favored allocation ratio is 1 : 3 when new offspring are full siblings. The difference occurs because the asymmetric

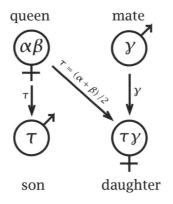

queen   mate

son   daughter

Figure 10.4   Calculation of inclusive fitness coefficients for a haplodiploid genetic system. Each letter is the average effect of an allele. The different letters are random variables drawn from the same population, that is; they are
alleles at the same genetic locus. The variance of each variable is the same,
and can be taken as standardized to one. The relatedness of an actor, a, to
a recipient, b, is r_{ba}, written in this way to emphasize that the relatedness
is a regression of recipient on actor. Specifically, relatedness was defined in
Eq. (3.9) as $r = \mathrm{Cov}(G, g)/\mathrm{Var}(g)$, where g is the actor's breeding value, and G
is the recipient's breeding value. In the diagram, breeding value is the average
value of the allelic effects. Thus the queen's breeding value is $\tau = (\alpha + \beta)/2$,
which is also the average value of a gamete she transmits to offspring. If we
define the correlation between any pair of the three parental alleles, α, β, γ, as
F, the inbreeding coefficient, then it is easy to calculate the relatedness coefficients by working out the appropriate covariances (see Section 3.5; Hamilton
1972, makes these same calculations based on pedigree analysis). The relatedness of a queen to a male (son) is $r_{mq} = 1$, and of a queen to a daughter,
$r_{fq} = (1 + 3F)/(2 + 2F)$. A worker is a female offspring of the queen. Her relatedness to a sister is $r_{fw} = (3 + 5F)/(4 + 4F)$, and her relatedness to a brother is
$r_{mw} = (1 + 3F)/(2 + 2F)$. These calculations all assume that the mother mates
only once, and that the average effects of alleles do not change between generations. With these coefficients, we have, from the queen's point of view, the
favored allocation ratio $c_m r_m : c_f r_f$ as $1 + F : 1 + 3F$. With no inbreeding, this
ratio is $1 : 1$ with no inbreeding. From the worker's point of view, the favored
ratio is $1 + 3F : 3 + 5F$, and, with no inbreeding, this ratio is $1 : 3$.

genetics of haplodiploidy cause relatively high relatedness of females to
their sisters (Fig. 10.4).

The different allocation ratios favored by the queen and her daughter-
workers cause conflict within colonies (Trivers and Hare 1976). Many
studies have measured allocation ratios to determine which of the following prevails: queen control, with $1 : 1$, or worker control, with $1 : 3$
(Crozier and Pamilo 1996). Several studies suggest worker control, but

there are many difficulties of measurement, uncontrolled factors, and interpretation.

The essential problem is that the prediction of worker control requires a fit: do the observed sex allocation ratios fit 1 : 3 better than 1 : 1? I have mentioned many times that fit is nearly impossible to interpret. It is much easier to test a comparative prediction. For example, how does one expect the sex allocation to change with some parameter? How does the pattern of change differ between queen and worker control?

Boomsma and Grafen (1990, 1991) developed an interesting comparative hypothesis that distinguishes between worker and queen control. They argue as follows. Suppose that the number of mates per queen varies among colonies. From the queen's point of view, her relatedness to sons and daughters is unaffected by her number of mates. Thus her favored allocation ratio, $r_{mq} : 2r_{fq}$, does not change with multiple mating. The subscript q in the relatedness coefficients emphasizes that these coefficients are taken from the point of view of the queen as actor.

From the workers' point of view, the favored ratio of male to female offspring, $r_{mw} : 2r_{fw}$, varies with the number of matings by the queen. The relatedness coefficients from the workers' point of view vary for the following reasons. The haploid males inherit only from the mother; thus the workers' relatedness to brothers is not affected by multiple mating. The diploid females inherit from both mother and father; thus the workers' relatedness to sisters depends on whether they are full or half siblings. With multiple mating, the probability that a female is a half sibling increases, which decreases r_{fw} and skews the workers favored sex ratio, $r_{mw} : 2r_{fw}$, toward males.

Suppose the workers can assess whether they are in a full-sibling or half-sibling colony. Then, under worker control, the distribution of colony sex allocation ratios is predicted to be bimodal. The full-sibling colonies are favored to produce relatively more females than do the mixed-sibling colonies because the workers value the females more highly when they are full siblings. If the queen is in control, the relatedness asymmetry from the workers' point of view has no influence on predicted sex ratios. A few studies suggest that sex ratios change with number of matings by the queen in the manner predicted under worker control (Boomsma and Grafen 1990; Sundström 1995; Sundström and Keller 1996).

This idea emphasizes the importance of studying variation among colonies. A complementary prediction about colony variation has also received preliminary support. In Frank (1987c), I analyzed the relative effects of local competition for mates among male relatives and local resource competition among female relatives. I showed that if local mate competition is stronger than local resource competition, smaller colonies are favored to invest a relatively greater portion of their resources in males than that invested by larger colonies. By contrast, stronger local resource competition than local mate competition favors small colonies to invest relatively more in female offspring than do large colonies (see *Simultaneous Game*, p. 200).

Sundström (1995) found that some ant colonies in her study followed the prediction of relatively stronger local resource competition. Hasegawa and Yamaguchi (1995) observed mating competition among male relatives and sex ratio variation among colonies matching the prediction of stronger local mate competition. However, several studies do not show the trends predicted by the local competition models, apparently because sex-biased competition among relatives does not occur in many species (Bourke and Franks 1995, 194–195).

SPLIT SEX RATIOS AND THE ORIGINS OF SOCIAL BEHAVIOR

Worker-queen conflict over sex allocation occurs in established colonies. Another problem concerns the evolutionary origin of workers: daughters who forgo their own reproduction to aid the reproduction of their mother. Suppose, for example, that we start with lone mothers who produce offspring without aid from daughters. What aspects of sex allocation could favor a daughter to stay with her mother and help to raise siblings, rather than to nest alone and raise her own offspring?

The relatedness asymmetries of haplodiploidy cause a female to gain more by raising her sisters than by raising her own offspring (Hamilton 1972; Trivers and Hare 1976). This can be shown by following through the details of relatedness, reproductive value, and marginal value. But the essential process can be illustrated by simply examining the class reproductive value and relatedness weightings for offspring, using the case of outbreeding and the coefficients in the caption of Fig. 10.4. From a mother's (queen's) point of view, her value for sons and daughters is

$$c_m r_{mq} = 1/2$$
$$c_f r_{fq} = 1/2,$$

where the relatedness coefficients use the q subscript to emphasize that they are taken from the queen's point of view, as in the caption of Fig. 10.4. From a sister's (worker's) point of view

$$c_m r_{mw} = 1/4$$
$$c_f r_{fw} = 3/4,$$

with the w subscript emphasizing the workers' point of view. Thus a female can gain more per offspring by working to raise sisters than by directly producing sons and daughters, but she gains less per brother than for an offspring of her own. Daughters are increasingly favored to help as the sex ratio of the mother's brood becomes more female-biased.

Factors that predictably cause relatively female-biased broods may have contributed to the origin of workers in some species (Trivers and Hare 1976). This has led to a few theories that match female-biased broods with worker behavior. I briefly mention four ideas.

The first workers may have directly manipulated the sex ratio of their mothers by killing male eggs or by investing more in female offspring. Two factors make this idea unlikely: the workers would have to be able to detect offspring sex early in development, and the reduction of brothers' fitnesses might not be sufficiently compensated by production of additional female eggs.

The second idea is Seger's (1983) analysis of overlapping generations. Some species produce two broods per year, with differential survival of males and females over the two generations. Seger showed that regular alternations in sex ratio are predicted in this life cycle. In some cases, mothers producing a second brood are favored to make a relatively female-biased sex ratio. This life history satisfies two requirements for the evolution of workers. First, older daughters are available when the mother produces her second brood, and could help their mother produce offspring. Second, the later broods are relatively female biased. Thus older daughters may gain more by raising sisters than by producing sons and daughters. The theory requires matrix analysis of reproductive value, which I develop in the next chapter.

The third model describes a synergism between sib-rearing and sex ratio (Frank and Crespi 1989). The hypothesis depends on three conditions. First, daughters that help cause more food to be provisioned per offspring, which in turn causes larger offspring. Second, females gain more than males by being large, which favors mothers with helpers to produce a higher proportion of daughters. Third, a female worker's fitness rises as her mother's brood becomes increasingly female biased because a female worker is more closely related to her sisters than to her brothers. Boomsma (1991) and Boomsma and Eickwort (1993) describe preliminary evidence supporting this model.

The final idea is an extension of the constant male hypothesis (see Fig. 9.1). Under local mate competition, mothers that produce relatively more offspring are favored to make extra offspring into daughters. If workers increase the number of offspring reared, then the mother is favored to make daughters for most of those additional offspring. This creates an association between sib-rearing and female-biased sex ratios. I am not aware of any previous discussion of this model in the context of worker behavior.

11

Sex Allocation: Reproductive Value

Sex allocation is the division of resources between male and female reproduction. The optimal split depends on the relative reproductive value of these two fitness components. Sex allocation may, in addition, influence survival, fecundity, and other components of fitness. An actor's division of resources into male and female can affect neighbors' fitnesses. The proper weighting of all these fitness components requires attention to reproductive value. In this chapter, I survey the main concepts and methods of analysis.

11.1 Current versus Future Reproduction

The prior sex allocation models analyzed the division of resources into male and female components. An actor may also reserve resources for the future.

The question is: does the split between current and future reproduction influence the ratio of resources devoted to males and females? The answer turns out to be no if returns on male and female investment are linear (Leigh 1970; Charlesworth 1977; Charnov 1982). However, the split between survival and reproduction does influence the sex allocation ratio when the relative slopes of male and female returns differ with scale. The models in this section clarify these conclusions.

I begin with a simple marginal value model. I assume that diploid mothers control sex allocation. The returns on investment are given by $\mu(x)$ for sons, and $\phi(y)$ for daughters, with total investment of $k = x+y$. If k is constant among mothers, then the usual analysis shows that the equilibrium in a randomly mating population, (x^*, y^*), is given by

$$\frac{\mu'(x^*)}{\mu(x^*)} = \frac{\phi'(y^*)}{\phi(y^*)}, \tag{11.1}$$

where μ' denotes differentiation with respect to x, and ϕ' denotes differentiation with respect to y.

Now suppose that a mother can enhance her own survival by withholding resources from current sons and daughters. Let a mother's survivorship to the next reproductive season be $\psi(z)$. Total allocation to sons, daughters, and survivorship sums to one, $x + y + z = 1$.

Current and future reproduction must be assigned proper weights for reproductive value, using the methods of Chapter 8. The first step is to describe the contribution of each class to other classes in a demographic matrix

$$\mathbf{A} = \begin{bmatrix} 0 & n\alpha & 0 & n \\ \mu(x) & t & 0 & 0 \\ 0 & n\alpha & 0 & n \\ 0 & 0 & \phi(y) & \psi(z) \end{bmatrix}, \tag{11.2}$$

where the columns and rows are, from first to last, juvenile males, adult males, juvenile females, and adult females. Thus a mother (last column) has $2n$ sons and $2n$ daughters, of which one-half are credited to her and one-half are credited to her mates. She survives to the following season with probability $\psi(z)$. Her juvenile sons survive to reproductive age with probability $\mu(x)$, given in the first column. In this formulation, $n\mu(x)$ can be thought of as the mother's fecundity through sons, or one can consider $\mu(x)$ as the survival a mother gives to each of her sons. Similarly, juvenile daughters survive to reproductive age with probability $\phi(y)$. An adult male expects credit for $n\alpha$ sons and $n\alpha$ daughters, where α is the number of mates per male, given by the ratio of the number of adult females to adult males in the population. Adult males survive each season with probability t.

Demographic properties of the population are influenced by the population growth rate, given by the largest eigenvalue of the normal matrix $\mathbf{A}^*$. For this step it is convenient to rewrite the matrix by assigning all juveniles to the adult females

$$\mathbf{F}^* = \begin{bmatrix} 0 & 0 & 0 & 2n \\ \mu(x^*) & t & 0 & 0 \\ 0 & 0 & 0 & 2n \\ 0 & 0 & \phi(y^*) & \psi(z^*) \end{bmatrix}.$$

The dominant eigenvalue is the largest value of λ that satisfies the characteristic equation

$$\lambda^2 - \lambda\psi^* - 2n\phi^* = 0,$$

where $\psi^* = \psi(z^*)$ and $\phi^* = \phi(y^*)$ are the functions evaluated at their equilibrium arguments.

The next step is to obtain the equilibrium class frequencies, $\mathbf{u} = (\hat{u}_0, \hat{u}_1, u_0, u_1)$, where $\mathbf{u}$ is a column vector, $\hat{u}$ and u denote male and female classes, respectively, and the subscripts 0 and 1 denote juvenile and adult, respectively. This calculation and the ones that follow all use the original matrix in Eq. (11.2), evaluated at x^*, y^* and z^*. The values for $\mathbf{u}$ are obtained from $\mathbf{A}^*\mathbf{u} = \lambda\mathbf{u}$, yielding

$$\hat{u}_0 = 2n/\lambda$$
$$\hat{u}_1 = (\lambda - \psi^*) / (\lambda - t)$$
$$u_0 = 2n/\lambda$$
$$u_1 = 1,$$

where u_1 is arbitrarily set to one, and the other values are given as number of individuals per adult female in the population.

The individual reproductive values are obtained from $\mathbf{v}\mathbf{A}^* = \lambda\mathbf{v}$, yielding

$$\hat{v}_0 = \phi^*/\lambda$$
$$\hat{v}_1 = \phi^*/\mu^*$$
$$v_0 = \phi^*/\lambda$$
$$v_1 = 1.$$

Each mother has three characters: investment in males, x, investment in females, y, and investment in survival, z. The constraint $x + y + z = 1$ reduces the number of characters to two, x and y, and the constrained character $z = 1 - x - y$.

The next step, by the standard methods of Chapter 8, is to write total recipient fitness as $W = \mathbf{v}\mathbf{A}\mathbf{u}$, differentiate with respect to transmitted breeding value for each of the two characters, and attempt to find a local equilibrium. I take g as breeding value for x, and h as breeding value for y. I also assume that under random mating, transmitted breeding values equal individual breeding values. I take the two traits as uncorrelated by setting $dx/dh = dy/dg = 0$.

The conditions $dW/dg = 0$ and $dW/dh = 0$ yield, respectively

$$\tilde{r}\,\hat{u}_0\hat{v}_1\mu'\,(x^*) - \psi'\,(1 - x^* - y^*) = 0$$
$$\tilde{r}\,u_0 v_1 \phi'\,(y^*) - \psi'\,(1 - x^* - y^*) = 0,$$

where primes denote differentiation with respect to the standard argument for each function, and $\tilde{r}$ is the coefficient of relatedness for direct fitness between offspring and mother. Replacing this coefficient with the inclusive fitness coefficient, r, and expanding the u and v terms yields

$$r\,(\lambda - \psi^*)\,\frac{\mu'}{\mu^*} - \psi' = 0 \qquad (11.3a)$$

$$r\,(\lambda - \psi^*)\,\frac{\phi'}{\phi^*} - \psi' = 0. \qquad (11.3b)$$

These equations show that the standard marginal value result holds

$$\frac{\mu'\,(x^*)}{\mu\,(x^*)} = \frac{\phi'\,(y^*)}{\phi\,(y^*)},$$

as in Eq. (11.1). Given this marginal value result, we can return to the original question. Does the split between current and future repro-duction influence the ratio of resources devoted to males and females, $x^* : y^*$? Briefly, the answer is yes, whenever the relative slopes of μ and ϕ differ with scale. A few examples clarify this conclusion.

The first example sets the relative slopes of male and female return independently of scale by using power functions

$$\mu\,(x) = x^{s_m}$$
$$\phi\,(y) = y^{s_f}$$
$$\psi\,(z) = tz^{s_z},$$

with $0 < s_m, s_f, s_z \leq 1$. The sex allocation ratio $x^* : y^*$ is given by $s_m : s_f$. If we set $s = s_m = s_f$ and $s_z = 1$, then the equilibrium allocations can be solved for directly

$$x^* = y^* = \frac{sr\,(\lambda - t)}{t\,(1 - 2sr)}$$

$$z^* = \frac{t - 2sr\lambda}{t\,(1 - 2sr)}.$$

Some general comparative predictions follow. First, as s_m and s_f de-cline, larger allocations to sons and daughters provide low marginal returns; thus allocation to survival increases. Second, as λ increases, population expansion enhances the reproductive value of offspring rel-ative to parents, which tends to reduce the value of allocation to sur-vival. Third, as t decreases, larger allocation to survival provides lower marginal returns, which reduce survival in favor of reproduction. Fi-nally, an increase in parent-offspring relatedness, r, favors greater allo-cation to reproduction.

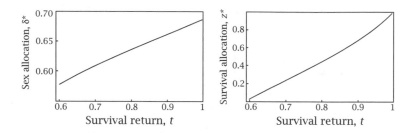

Figure 11.1   Change in the sex allocation ratio with a change in the split be-tween investment in reproduction and investment in survival. The sex allocation ratio is the fraction of reproductive resources invested in males, $\delta = x/(x+y)$. The solution is based on the conditions in Eqs. (11.3), with functional forms given in Eqs. (11.4). Parameters are $\alpha = 0.1$, $s = 0.5$, $r = 0.5$, and $\lambda = 1$.

The second example compares male and female return functions in which the ratio of slopes differs with scale. It is convenient to give the derivatives of the return functions

$$\mu'(x) = x^\alpha (1 - x) \tag{11.4a}$$

$$\phi'(y) = sy^{s-1} \tag{11.4b}$$

$$\psi'(z) = t, \tag{11.4c}$$

where the constants of integration are zero. A numerical example, based on the equilibrium conditions in Eqs. (11.3), is shown in Fig. 11.1. The fraction of reproductive allocation given to males is δ^*. In this case, greater allocation to survival is associated with an increase in the relative allocation to males.

11.2 Shifts in Sex Allocation with Age

Reproductive allocation may increase with age for a variety of reasons (Charnov 1982; Charlesworth 1994).   Older individuals may be larger and have more resources available. Or, the probability of survival may decrease with age (senescence), favoring an increase in reproductive al-location. When the relative slopes of male and female returns differ with scale, then shifts in total reproductive allocation cause shifts in the proportion of resources invested in males and females.

I present a model in which mating is random, and mothers die af-ter the Nth reproductive season. In prior years, a mother divides re-sources among sons, daughters, and survival. This simplified form of senescence may favor a shift with age in relative allocations to males, females and survival (Charnov 1982; Charlesworth 1994). I illustrate the methods of analysis. Modified assumptions will be required for many applications.

I begin with the transition matrix

$$
\mathbf{A} = \begin{bmatrix}
m_0 & m_1 & f_0 & f_1 & f_2 & \cdot & f_N \\
0 & M/2\hat{u}_1 & 0 & n\mu(x_1) & n\mu(x_2) & \cdot & n\mu(x_N) \\
k & t & 0 & 0 & 0 & \cdot & 0 \\
0 & F/2\hat{u}_1 & 0 & n\phi(y_1) & n\phi(y_2) & \cdot & n\phi(y_N) \\
0 & 0 & k & 0 & 0 & \cdot & 0 \\
0 & 0 & 0 & \psi_1 & 0 & \cdot & 0 \\
0 & 0 & 0 & 0 & \psi_2 & \cdot & 0 \\
\cdot & \cdot & \cdot & \cdot & \cdot & \cdot & 0 \\
0 & 0 & 0 & 0 & 0 & \psi_{N-1} & 0
\end{bmatrix}.
$$

The top row defines each column. The first two, m_0 and m_1, are juvenile and adult males, respectively. The f_i are females of age i. The survival probability of juveniles to the adult stage is k. Adult males survive each season with probability t. Each mother of age $i > 0$ produces $2n\mu(x_i)$ males and $2n\phi(y_i)$ females, of which one-half are credited to the mother and one-half to the father. The total numbers of male and female juveniles in each generation are proportional to

$$
M = 2n \sum_i \mu(x_i^*) u_i
$$

$$
F = 2n \sum_i \phi(y_i^*) u_i,
$$

where u_i is the number of adult females of age i per juvenile female (see below). Each adult male is credited with $M/2\hat{u}_1$ sons and $F/2\hat{u}_1$ daughters, where $\hat{u}_1$ is proportional to the total number of adult males. Mothers of age $i < N$ survive to the next breeding season with probability $\psi_i = \psi(z_i)$. All mothers die after the Nth season.

The matrix can be rearranged by assigning all progeny to the mothers. Thus, a mother of age i produces $2n\mu(x_i)$ sons and $2n\phi(y_i)$ daughters, and adult males are assigned values of zero for contribution to juvenile classes. The female-dominant matrix allows easy calculation of the characteristic equation for population growth (Charlesworth 1994), yielding

$$
\lambda = 2n \sum_{i=1}^{N} \phi_i u_i = F,
$$

where $\phi_i = \phi(y_i^*)$, and the definition is given below for u_i, the number of females of age i. I will also use the abbreviations $\mu_i = \mu(x_i^*)$ and $\psi_i = \psi(z_i^*)$.

The next step is to calculate, for each class, the number of individuals per juvenile female

$$\hat{u}_0 = M/F$$

$$\hat{u}_1 = k\hat{u}_0/(\lambda - t)$$

$$u_0 = 1$$

$$u_i = \prod_{j=1}^{i} \psi_{j-1}/\lambda^i,$$

where $\psi_0 = k$.

The individual reproductive values for each class, relative to a juvenile female, are

$$\hat{v}_0 = F/M$$

$$\hat{v}_1 = \hat{v}_0 (\lambda/k)$$

$$v_0 = 1$$

$$\lambda v_i = n \sum_{j=i}^{N} [(\hat{v}_0 \mu_i + v_0 \phi_i)] \prod_{s=i}^{j-1} (\psi_s/\lambda).$$

The solution must be given in terms of two allocation decisions at each age: the mother's split between reproduction and survival, and her division of reproductive allocation between sons and daughters. Thus we seek the equilibrium trait values (x_i^*, y_i^*, z_i^*) for $i = 1, \ldots, N$. I assume that the trait values for each time period are independent of past and future trait values. If we focus on the variable traits in each time period, then from the transition matrix $\mathbf{A}$, the recipient fitnesses influenced by characters at time i are proportional to

$$W_i = \hat{v}_0 n \mu (x_i) + v_0 n \phi (y_i) + v_{i+1} \psi (z_i),$$

where $v_{N+1} = 0$.

Total allocation is, of course, limited in each time period. Two types of constraint must be distinguished. On the one hand, total resources allocated to both survival and reproduction are limited. On the other hand, given a fixed amount allocated to reproduction, there is a tradeoff between male and female allocation.

The division between survival and reproduction can be expressed, for example, by the constraint on allocation to survival, $z_i = 1 - \gamma x_i - y_i$

for all time periods except the last. Here male and female investment have different effects on survival according to the parameter y. Death is imminent in the Nth time period; thus allocation to survival is zero, and the constraint concerns the split between male and female allocation, $\beta x_N + y_N = 1$.

The usual methods yield two sets of conditions for candidate equilibria. The first describes the split between male and female allocation in each time period

$$\frac{\mu_i'}{M} = \frac{y\phi_i'}{F} \qquad i = 1, \ldots, N-1$$

$$\frac{\mu_i'}{M} = \frac{\beta\phi_i'}{F} \qquad i = N.$$

This is the standard marginal value result for sex allocation, with the addition of the y and β factors. The term y is a linear scaling for the cost to survival per unit of male and female allocation, from the constraint $z_i = 1 - yx_i - y_i$. The term β is a linear scaling factor for the tradeoff between male and female allocation in the Nth period, from the constraint $\beta x_N + y_N = 1$. If male allocation is more costly to survival than to fecundity, $y > \beta$, then mothers will tend to shift their allocation toward males later in life, as the marginal gains of allocation to survival decline (Charnov 1982; Charlesworth 1994). The argument can, of course, be generalized for arbitrary constraints to account for the partial effect of male allocation on female allocation, holding constant allocation to survival. In the general case, the scaling factors are $-\partial y_i/\partial x_i$.

The second set of equilibrium conditions determines the split between allocation to current reproduction and survival

$$rn\phi_i' = \delta v_{i+1}\psi_i' \qquad i = 1, \ldots, N-1.$$

The left side is the marginal return on investment in daughters, weighted by the mother's relatedness to daughters, r. The right side is the marginal return for shifting a unit of investment from daughters to survival, where $\delta = -dz_i/dy_i$ is the scaling between investment in daughters and investment in survival. The term ψ' is the marginal change in survival probability, and the term v_{i+1} is the value of surviving to the next time period.

Suppose that returns on additional investment in daughters increase at a diminishing rate, $\phi_i'' < 0$. As a mother ages, her future reproductive value, v_{i+1}, declines. Thus an aging mother is favored to increase her allocation to daughters until marginal returns decline to match marginal gains for future reproductive value via survival. An increased allocation to daughters may be associated with a change in the split between daughters and sons. This split is determined by the marginal gains for additional investment in each sex given in the first set of equilibrium conditions.

11.3 Perturbation of Stable Age Structure

Fisherian allocation of one-half to each sex holds in the prior models when returns are equivalent for male and female investment. Those models assumed stable age structure. Werren and Taylor (1984) showed that a perturbation of the stable age structure shifts the favored allocation ratio toward the sex with the lower variation of age-specific fitnesses. The shift occurs because an increase in age-specific fitness follows a diminishing-return curve; thus lower variation across years provides higher lifetime fitness.

Let age-specific fitness, p_i, be the contribution of individuals of age i to newborns in the current year, with $\sum p_i = 1$. Suppose exceptional recruitment of newborns occurs in a particular year. The size of this cohort is $1 + \alpha$ relative to a typical cohort. Such perturbations are sufficiently rare that, among living age classes, there is never more than one age class of exceptional size.

The reproductive value of an individual born into an exceptional age class is

$$v = \sum_i \frac{p_i}{1 + \alpha p_i}.$$

The denominator accounts for the increased number of competitors added by the exceptional cohort.

The effect of exceptional recruitment can be studied by considering the properties of v. When α is small, the return for investment in the exceptional cohort in each year is $\beta(p) = p/(1 + \alpha p) \approx p(1 - \alpha p)$. Thus return for investment increases at a diminishing rate as p increases.

The total return over the n age classes can be written as n multiplied by the expected return in each year

$$v = nE[\beta(p)].$$

We can think of p as an intermediate measure of success in a sequence of investments. The investment is to add an individual to the exceptional cohort at each age. The intermediate measure of success is the age-specific fitness at each age, p. The actual return at each age is $\beta(p)$. The total return over all ages is v.

The problem can be restated as follows. How can one relate variation in an intermediate measure of success, p, to the expected value of total success? The standard approach from economic theory is to expand $\beta(p)$ by the Taylor series (for behavioral applications in biology, see Real 1980; Stephens and Krebs 1986; Frank and Slatkin 1990b), yielding

$$E[\beta(p)] = \beta(\overline{p}) + \beta''(\overline{p})\,\sigma_p^2/2 + \epsilon,$$

where σ_p^2 is the variance in p, the second derivative of β is β'', and ϵ is the remainder term.

When α is small we have, from above, $\beta(p) \approx p(1 - \alpha p)$. Using the fact that $\overline{p} = 1/n$, the reproductive value of an individual added to the exceptional cohort is

$$v \approx 1 - \alpha/n - n\alpha\sigma_p^2.$$

The favored sex allocation in the exceptional cohort may deviate from the norm when the age-specific fitnesses differ between the sexes. Consider a diploid population, with linear returns on investment in each sex. Then Fisher's theory holds, and the allocation ratio in a stable population is one-half to each sex. Let the allocation ratio during the year of exceptional recruitment be $x = (1/2)(1 + \delta)$, where x is the fraction of resources invested in males.

The reproductive value of an extra male added to the exceptional cohort is

$$v_m = \sum_i \frac{m_i}{1 + \alpha m_i + \delta(1 + \alpha)\,m_i},$$

where m_i is the age-specific fitness of males, and $\delta(1 + \alpha)$ is the proportional increase in the number of males in the exceptional cohort as the allocation ratio, x, changes from one-half. If, as above, α is small, then the deviation in the allocation ratio, δ, will also be small, and we can ignore terms of order $\delta\alpha$. Thus

$$v_m \approx \sum_i \frac{m_i}{1 + (\alpha + \delta)\,m_i}$$

$$\approx 1 - (\alpha + \delta)/n - n(\alpha + \delta)\,\sigma_m^2,$$

where σ_m^2 is the variance in age-specific fitnesses of males. A similar calculation for females yields

$$v_f \approx 1 - (\alpha - \delta)/n - n(\alpha - \delta)\,\sigma_f^2.$$

The equilibrium sex allocation for a diploid population with linear returns on each sex occurs when $v_m = v_f$. The equilibrium allocation ratio, expressed as δ, the proportional increase in the fraction of males relative to one-half, is

$$\delta = \frac{\alpha n^2 \left(\sigma_f^2 - \sigma_m^2 \right)}{2 + n^2 \left(\sigma_f^2 + \sigma_m^2 \right)},$$

showing that the allocation ratio is biased toward the sex with lower variance in age-specific fitness. This is equivalent to the result given by Werren and Taylor (1984). West and Godfray (1997) have shown that allocation ratios in cohorts after the exceptional year are also favored to deviate from the norm. These later deviations influence the allocation ratio in the exceptional year.

11.4 Cyclical Age Structure with Male–Female Asymmetry

Asymmetry in sex-specific life schedules can bias the reproductive values of the sexes. Consider, for example, Seger's (1983) model of insect life history (extending Werren and Charnov 1978). There are two generations per year, spring and summer. Males and females are born in the spring, mate, and reproduce. The females die, and a portion of the males survives to mate again in the summer generation. Some offspring from the spring matings emerge in the summer generation; others overwinter in a pre-adult form and emerge the following spring. All offspring from the summer mothers emerge in the spring.

Males born in the spring reproduce in both the spring and summer generations. By contrast, males born in the summer reproduce only in the generation in which they are born. Thus the reproductive value of the spring males is higher than that of the summer males, and consequently the sex ratio is biased toward males in the spring and females in the summer. I recast Seger's (1983) analysis into the framework presented in this book; Grafen (1986) provided a similar method based on reproductive value.

ALTERNATIVE DEMOGRAPHIC MATRICES

Complex demographies do not present any conceptual difficulties for the standard method. The fitness matrices and expressions for reproductive value can, however, be complex. It is often convenient to work with alternative forms of the standard matrix of individual fitnesses. This matrix is $A = [w_{ij}]$, where w_{ij} is the contribution of an individual of class j to class i in the following time period. The abundance of each class is u_j in our standard notation. Thus the class fitness matrix is $B = [w_{ij}u_j]$, where each entry is the total contribution of class j to class i. Abundances in the next time period are $u' = Au = B1$, where 1 is the $n \times 1$ vector of ones.

The definition of total recipient fitness in the theory is $W = vAu$, where each v_j is the reproductive value of an individual of class j (see Eq. (8.2)). We can also write total fitness as $W = vB1$.

Finally, a matrix based on female reproduction is useful for studying class abundances and population growth. This matrix is obtained by starting with B^*, the normal matrix of class fitnesses. Next, assign all newborn offspring entirely to mothers, creating a normal matrix D^*. This does not change the numbers in any class if female fecundity is independent of the numbers of males. The abundances in the next time period are $u' = D^*1$. Thus D^* is a useful matrix to define aspects of population dynamics and the abundances of classes.

ABUNDANCE

The population biology of this model is given by

$$D^* = \begin{bmatrix} m_0 & m_1 & f_0 & f_1 \\ 0 & 0 & t\hat{u}_0 & (1-t)\,\hat{u}_0 \\ s\hat{u}_0 & 0 & \omega x_1^* & 0 \\ 0 & 0 & t & 1-t \\ 0 & 0 & \omega\,(1-x_1^*) & 0 \end{bmatrix},$$

where the first row gives the class definitions, m_0 and m_1 for spring and summer males, and f_0 and f_1 for spring and summer females. In this D^* matrix, all newborns are assigned to the mother and entries give the total contributions of each class to other classes. As noted above, abundances in the next time period are given by $u' = D^*1$. Put another way, the abundance of class i is the sum of entries in the matrix across row i. I use the notation $\mathbf{u} = (\hat{u}_0, \hat{u}_1, u_0, u_1)$ for class abundances, with the hats denoting males. I normalize abundances so that the number of females in the spring generation is $u_0 = 1$.

Consider the third row, the contribution of other classes to female offspring in the spring generation. A fraction t of the spring females comes from the prior spring, and a fraction $1 - t$ of the spring females comes from the prior summer. The total number of spring females is regulated to a constant multiple of $u_0 = t + (1 - t) = 1$.

The fourth row is the number of summer females. These are produced only by spring mothers. Each spring mother has ω surviving progeny, of which $1 - x_1^*$ are daughters. Here, x_1 is the fraction of progeny that are males when contributed to the summer generation, and x_0 is the fraction of progeny that are males when contributed to the spring generation. The total number of summer females is $u_1 = \omega(1 - x_1^*)$.

The first row is the number of spring males. A fraction t comes from the prior spring, and a fraction $1 - t$ comes from the prior summer. The class abundance, $\hat{u}_0 = x_0^*/(1 - x_0^*)$, is the number of males per female in the spring generation. Because the total number of females in the spring is regulated to one, the total number of males in the spring is $\hat{u}_0$.

The second row is the number of summer males. A fraction s of the $\hat{u}_0$ males survives from the spring to the summer. The spring females produce ωx_1^* sons. Thus the abundance of summer males is $\hat{u}_1 = s\hat{u}_0 + \omega x_1^*$.

From these definitions, it is clear that class abundances are stable, $\mathbf{u}' = \mathbf{D}^*\mathbf{1} = \mathbf{u}$. Thus the dominant eigenvalue is $\lambda = 1$. Stability arises because the number of females born in the spring is regulated to a constant, and all adult individuals die after the summer. The dominant eigenvalue can also be calculated in the usual way, first recovering the standard fitness matrix $\mathbf{A}^*$, and then solving the characteristic equation.

In summary, the stable class abundances are

$$\hat{u}_0 = x_0^*/(1 - x_0^*)$$
$$\hat{u}_1 = s\hat{u}_0 + \omega x_1^*$$
$$u_0 = 1$$
$$u_1 = \omega(1 - x_1^*).$$

Reproductive Value

The normal matrix, $\mathbf{D}^*$, assigns all progeny to mothers. To obtain reproductive values, we need the gametic contributions of males and females. I assume a haplodiploid genetic system, to match the application of this model to haplodiploid bees and wasps (Seger 1983). Under haplodiploidy, males do not contribute to male progeny, and thus male offspring are credited entirely to mothers. The female progeny are split equally between male and female parents. This division leads immediately to the normal matrix, $\mathbf{B}^*$, the gametic contributions from each class

$$
\mathbf{B}^* = \begin{bmatrix} 0 & 0 & \frac{tx_0^*}{(1-x_0^*)} & \frac{(1-t)x_0^*}{(1-x_0^*)} \\ \frac{sx_0^*}{1-x_0^*} & 0 & \omega x_1^* & 0 \\ \frac{t}{2} & \frac{1-t}{2} & \frac{t}{2} & \frac{1-t}{2} \\ \frac{\omega(1-x_1^*)}{2} & 0 & \frac{\omega(1-x_1^*)}{2} & 0 \end{bmatrix}.
$$

The individual fitness matrix, $\mathbf{A}$, is obtained from the definitions above, $\mathbf{B} = [w_{ij}u_j]$ and $\mathbf{A} = [w_{ij}]$. Individual reproductive values are obtained from $\mathbf{vA}^* = \lambda \mathbf{v}$, where in this model $\lambda = 1$. Normalizing the reproductive value of spring females to one yields

$$
\hat{v}_0 = s\hat{v}_1 + \frac{t}{2\hat{u}_0}v_0 + \frac{\omega\left(1-x_1^*\right)}{2\hat{u}_0}v_1
$$

$$
\hat{v}_1 = \frac{1-t}{2\hat{u}_1}v_0
$$

$$
v_0 = 1
$$

$$
v_1 = \frac{1-t}{u_1}\left(\hat{v}_0\hat{u}_0 + v_0/2\right).
$$

These valuations can be read in a meaningful way. For example, the value of a spring male, $\hat{v}_0$, has three components. His value by surviving to the summer is the probability of survival, s, weighted by the value of summer males, $\hat{v}_1$. In the second term, value for contribution to females of the following spring is the proportion of next spring's females from the current spring, t, which is split among the $\hat{u}_0$ males. This term is divided by 2 for the male's gametic contribution, and weighted by the reproductive value of next spring's females, v_0. Finally, the spring generation produces a total of $\omega(1-x_1^*)$ summer females, one-half of which are credited to the spring males. Each spring male will be the father of $1/\hat{u}_0$ of the summer females. The weighting for the summer females is v_1.

An explicit solution for individual values can be obtained by rewriting the reproductive value equations above as

$$\hat{v}_0 = \frac{2s}{1+t}\hat{v}_1 + \frac{1}{2\hat{u}_0}$$

$$\hat{v}_1 = \frac{1-t}{2\hat{u}_1}$$

$$v_0 = 1$$

$$v_1 = \frac{1-t}{u_1}\left(\hat{v}_0\hat{u}_0 + 1/2\right).$$

FITNESS

The standard method of solution is to write total fitness as $W = \mathbf{vAu}$ and differentiate with respect to transmitted breeding value. As mentioned above, an equivalent form of total fitness is $W = \mathbf{vB1}$, where $\mathbf{B}$ is the fitness matrix of total class contributions. The normal matrix $\mathbf{B}^*$ is transformed into the variant fitness matrix $\mathbf{B}$ by making explicit how variant phenotypes of actors influence recipient fitness. I assume that mothers control the sex ratio and that mating is random. Thus the fitness matrix is

$$\mathbf{B} = \begin{bmatrix} 0 & 0 & \frac{tx_0}{(1-x_0^*)} & \frac{(1-t)x_0}{(1-x_0^*)} \\ \frac{sx_0^*}{1-x_0^*} & 0 & \omega x_1 & 0 \\ \frac{t}{2} & \frac{1-t}{2} & \frac{t(1-x_0)}{2(1-x_0^*)} & \frac{(1-t)(1-x_0)}{2(1-x_0^*)} \\ \frac{\omega(1-x_1^*)}{2} & 0 & \frac{\omega(1-x_1)}{2} & 0 \end{bmatrix},$$

where unstarred values of sex ratios in the third and fourth columns are the variant behaviors. There are two phenotypes, x_0 and x_1, controlled, respectively, by breeding values g and h. I assume that the traits are uncorrelated in the sense that cross-derivatives with respect to transmitted breeding value are zero, $dx_1/dg' = dx_0/dh' = 0$.

Candidate equilibria are obtained by simultaneous solution of

$$dW/dg' = dW/dh' = 0.$$

This yields the two conditions

$$\hat{v}_0\hat{r} = v_0 r$$

$$\hat{v}_1\hat{r} = v_1 r,$$

where I have used inclusive fitness coefficients $\hat{r}$ and r for relatedness of mothers to sons and daughters, respectively. In randomly mating haplodiploids, the coefficients are $\hat{r} = 1$ and $r = 1/2$ (see Section 10.4). These conditions lead to the solution, given in the same form presented by Seger (1983) for his "sphecid" model

$$x_1^* = \frac{3 - 4x_0^* - t}{2(2 - 2x_0^* - t)} \tag{11.5}$$

$$\frac{s}{\omega} = \frac{(1 - x_0^*)(2x_0^* - 1)(1 + t)}{2x_0^*(2 - sx_0^* - t)}.$$

An alternative form for the first equation clarifies the result

$$x_1^* = \frac{1}{2} - \frac{(s/\omega)x_0^*}{(1 - x_0^*)(1 + t)}.$$

If the spring males do not survive to the summer, $s = 0$, there is no male-female asymmetry, and the summer sex ratio is $x_1^* = 1/2$. From Eq. (11.5), one can also show that $x_0^* = 1/2$.

The proportion of spring males that compete in the summer mating season is s/ω. An increase in this ratio decreases the reproductive value of males born in the summer. The proportion of males born in the summer, x_1^*, therefore declines with a rise in s/ω. This decline in the birth of summer males is matched by a increase in the proportion of males born in the spring, x_0^*.

In summary, the sex ratio of offspring born in the spring is biased toward males. The sex ratio of offspring born in the summer is biased toward females. Mothers of the summer generation contribute only to the following spring. These summer mothers produce male-biased sex ratios. Mothers in the spring generation contribute to both the following summer and spring. These spring mothers make a mixture of male-biased and female-biased broods.

Alternative Life Histories and Biological Consequences

The life history above is based on the biology of sphecid wasps. Seger (1983) also studied an alternative life cycle based on halictid bees. In the halictid model, three distinct generations of adult females occur in each year. The spring adults are born and mate in the fall, then survive the winter to reproduce in the spring. The spring mothers each produce ω

offspring that mature and reproduce in the summer. The spring mothers also produce some progeny that do not mature until the autumn. Of the autumn adults, a fraction t comes from the spring mothers and a fraction $1 - t$ comes from the summer mothers.

The males do not survive the winter; thus there are no males in the spring season. The spring mothers, who mated in the fall, produce both males and females for the summer generation. The summer males mate with the summer females. A fraction s of the summer males survives to mate with the autumn females. These surviving males from the summer compete with males that mature in the autumn.

The biology is summarized in a **B** matrix, which gives the total contribution of each class to other classes

$$
\mathbf{B} = \begin{bmatrix}
m_1 & m_2 & f_0 & f_1 & f_2 \\
0 & 0 & \omega x_1 & 0 & 0 \\
s\hat{u}_1 & 0 & \frac{tx_2}{1-x_2^*} & \frac{(1-t)x_2}{1-x_2^*} & 0 \\
0 & 0 & 0 & 0 & 1 \\
0 & \frac{\omega(1-x_1^*)}{2} & \frac{\omega(1-x_1)}{2} & 0 & 0 \\
\frac{1-t}{2} & \frac{t}{2} & \frac{t(1-x_2)}{2(1-x_2^*)} & \frac{(1-t)(1-x_2)}{2(1-x_2^*)} & 0
\end{bmatrix}.
$$

The number of adult females in the autumn is normalized to one. The top row defines the classes, with subscripts 0, 1 and 2 for the spring, summer, and autumn generations, respectively. The sex ratios x_1 and x_2 are for the fraction of males produced in the summer and fall generations. Solving for the equilibria is, as always, purely mechanical once the biology has been clearly specified by a fitness matrix. Seger (1983) gives the explicit solution. Here I simply summarize Seger's conclusions.

The generation 1 (summer) males survive with probability s to mate again in the fall. Thus generation 1 males have added reproductive value, and the sex ratio is biased toward males in the summer generation, $x_1^* > 1/2$. The surviving summer males reduce the reproductive value of males born in the autumn, causing a female-biased autumnal sex ratio, $x_2^* < 1/2$.

A few spring generation females may survive to reproduce again in the summer. Their daughters, born in the summer, could either reproduce by themselves or help their mothers raise siblings. Summer mothers, who contribute to the fall generation, are favored to produce a female-biased sex ratio, $x_2^* < 1/2$. A summer daughter, if she nested alone, would raise mostly her own daughters. If she helped her mother, she would raise mostly sisters. A haplodiploid female who mates randomly is related to her daughters by $1/2$. She is, by contrast, related to her sisters by $3/4$ if her mother does not remate (see Section 10.4). Thus, if her contributions to reproduction have the same value to siblings or progeny, then she may be favored to help her mother rather than to nest alone.

The same potential exists in the sphecid model for daughters to help their mothers. In that case, the spring mothers (generation 0) may survive to reproduce again in the summer (generation 1). Daughters that mature in the summer could help their mothers raise offspring for the following spring. The mother's favored sex ratio for the spring, x_0^*, is, however, male biased in that model. A daughter that helps her mother would be raising mostly brothers. A haplodiploid female is more closely related to her sons than to her brothers (Section 10.4). Thus the sphecid life history favors a female to nest alone rather than to aid her mother.

11.5 Transmission of Individual Quality

Variation in individual quality can affect sex expression. For example, when males compete directly for mating, large males often gain a disproportionate share of reproductive success. Individuals of high quality are therefore favored to be male, and those of low quality are favored to be female (Trivers and Willard 1973). I discussed the theory of conditional sex expression in Section 9.3.

The biological assumptions vary from case to case. But it is convenient to describe the problem from a mother's point of view with respect to allocation to sons and daughters. A mother has quality level k, which influences the quality of offspring she can produce. The quality level varies among mothers.

The problem is typically formulated by writing reproductive returns as a function of investment. The returns on investing y in males may be written as $\mu(y)$. Similarly, the returns on investing z in females may be written as $\phi(z)$.

The quality level, k, is a limited resource that can be split between sons and daughters. A mother with resource level k must therefore split her resource into a fraction, x_k, given to sons, and $1 - x_k$ given to daughters. The functions μ and ϕ are offspring viability or fecundity.

The allocation problem is to find x_k^* by comparing properly normalized values of $\mu(kx_k)$ and $\phi[k(1 - x_k)]$, as outlined in Section 9.3.

Leimar (1996) pointed out that, in some cases, quality is a nondivisible property that affects all offspring. For example, social status of the mother may provide a benefit to all offspring, or the quality of the family territory may equally affect all offspring.

Nondivisible quality does not, by itself, change the main conclusions. A high-quality mother is favored to produce more of the sex that gains the most from a quality environment.

Leimar's (1996) contribution concerns the inheritance of quality. Suppose that a mother with high social status has two effects on her progeny. First, her status provides more resources to her offspring. Second, status is transmitted to daughters but not to sons.

These two effects may favor opposite patterns of sex allocation. For example, additional resources to offspring may provide higher fecundity to sons than to daughters. This difference occurs because with direct male competition, a few large males may obtain most of the matings in a group. In terms of offspring fitness, high-quality mothers are favored to produce sons.

But simply counting the fitness of sons and daughters is not sufficient. The sons will have progeny of average quality, that is, of average reproductive value. The daughters, who inherit maternal quality, will themselves have offspring of high quality. Thus daughters produce offspring of relatively higher reproductive value. The fitnesses of sons and daughters must be weighted by these differences in reproductive value.

Leimar (1996) provided an excellent analytical treatment of this problem. He used the matrix methods of reproductive value (Taylor 1990; McNamara 1991) that I have developed in Chapter 8. Here I outline one simple problem to clarify the assumptions. I then show the result obtained by Leimar.

In this case there are only two quality levels, high and low. A mother produces sons and daughters with the same quality level as her own with probability α and offspring of opposite quality with probability $1 - \alpha$. Matrilineal inheritance of quality occurs with $\alpha > 1/2$. For values of α just above one-half, transmission of quality is weak. As $\alpha \to 1$, the fidelity of inheritance increases to perfection. Quality affects a son's mating success, but he does not transmit his quality to offspring.

Let μ_1 be the mating success of a high-quality son, and μ_0 be the success of a low quality son. The ratio $\mu = \mu_1/\mu_0$ is the improvement in the reproductive success of sons attributable to an increase in quality. Similarly, let ϕ_1 be the fecundity of a high-quality daughter, and ϕ_0 be the fecundity of a low-quality daughter. The ratio $\phi = \phi_1/\phi_0$ is the improvement in the reproductive success of daughters attributable to an increase in quality. Assume $\mu > \phi$; that is, assume an increase in quality provides greater benefit to sons than to daughters. This factor favors high-quality mothers to produce sons.

High-quality mothers produce a fraction x_1 of sons among their progeny. Low-quality mothers produce a fraction of x_0 sons. The problem is to measure how the reproductive advantage of high-quality sons, $\mu > \phi$, compares with the advantage that daughters gain by matrilineal inheritance of quality.

The assumptions are, as always, fully determined by the fitness matrix **A**. Once this matrix is specified, the analysis is mechanical, although perhaps difficult technically. It is convenient to split the matrix into the contribution of male parents, $\mathbf{A}_m$, and the contribution of female parents, $\mathbf{A}_f$, with $\mathbf{A} = [\mathbf{A}_m \ \mathbf{A}_f]$. First, the contribution of female parents of low and high quality, f_0 and f_1, is

$$
\mathbf{A}_f = \frac{1}{2}
\begin{bmatrix}
f_0 & f_1 \\
\alpha\phi_0 x_0 & (1-\alpha)\,\phi_1 x_1 \\
(1-\alpha)\,\phi_0 x_0 & \alpha\phi_1 x_1 \\
\alpha\phi_0\,(1-x_0) & (1-\alpha)\,\phi_1\,(1-x_1) \\
(1-\alpha)\,\phi_0\,(1-x_0) & \alpha\phi_1\,(1-x_1)
\end{bmatrix},
$$

where the recipients are, in the order of the rows, sons of low and high quality, and daughters of low and high quality. The model is diploid, and the contribution of each parent is weighted by one-half.

The male half of the matrix is

$$
\mathbf{A}_m = \frac{1}{2U}
\begin{bmatrix}
m_0 & m_1 \\
\mu_0 M_0 & \mu_1 M_0 \\
\mu_0 M_1 & \mu_1 M_1 \\
\mu_0 F_0 & \mu_1 F_0 \\
\mu_0 F_1 & \mu_1 F_1
\end{bmatrix},
$$

where U is the total mating success of all males; thus μ_i/U is the fraction of the total mating success by a male of quality i. The terms M_i and F_i

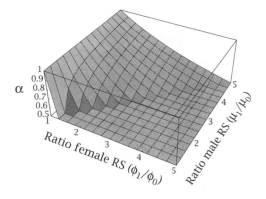

Figure 11.2 Conditions under which high-quality females are favored to produce daughters. The surface shows the minimum value of α required to favor an excess of daughters among high-quality females, from Eq. (11.6). The parameter $\alpha - 1/2$ is a measure of matrilineal inheritance. The term μ_1/μ_0 is the ratio of reproductive success (RS) for high-quality relative to low-quality males. The term ϕ_1/ϕ_0 is the same ratio for females.

are the total number of male and female progeny of quality i produced in the population. The explicit definitions for these terms are

$$U = \mu_0 \hat{u}_0 + \mu_1 \hat{u}_1$$
$$M_0 = \alpha \phi_0 x_0^* u_0 + (1 - \alpha)\, \phi_1 x_1^* u_1$$
$$M_1 = (1 - \alpha)\, f_0 x_0^* u_0 + \alpha \phi_1 x_1^* u_1$$
$$F_0 = \alpha \phi_0 \left(1 - x_0^*\right) u_0 + (1 - \alpha)\, \phi_1 \left(1 - x_1^*\right) u_1$$
$$F_1 = (1 - \alpha)\, \phi_0 \left(1 - x_0^*\right) u_0 + \alpha \phi_1 \left(1 - x_1^*\right) u_1,$$

where $\mathbf{u} = (\hat{u}_0, \hat{u}_1, u_0, u_1)$ are the numbers of males and females of quality 0 and 1.

Leimar (1996) proceeded by finding the vector of reproductive values, $\mathbf{v}$, and then calculating the conditions under which high-quality mothers are favored to produce more daughters than sons, $x_1^* < 1/2$. High-quality mothers are favored to produce an excess of daughters only when matrilineal inheritance, defined by α, is sufficient to offset the fecundity advantage of high-quality sons relative to high-quality daughters. The fecundity advantage is given by above by $\mu > \phi$.

Leimar (1996, Eq. 8) provided the condition on matrilineal inheritance, α, required to offset the male fecundity advantage. When the following condition is satisfied, high-quality mothers are favored to produce an excess of daughters

$$\alpha > \frac{1}{2} + \frac{(\mu - \phi)(\mu + 1)}{2(\mu + \phi)(\mu - 1)}. \tag{11.6}$$

The result is illustrated in Fig. 11.2.

11.6 Juveniles of One Sex Help Parents

Offspring sometimes remain with their parents to help raise siblings. Helpers may stay through the early part of their lives, and then reproduce when older. Or the helpers may be a sterile caste, never reproducing on their own.

In several bird species, the helpers are mostly of one sex. The helpers typically remain during their first few years of life, and attempt to reproduce later. The helpers in many species are male, and the adult sex ratio is often male biased (Brown 1978; Emlen 1978, 1984). In red-cockaded woodpeckers, the juvenile males help, and the juvenile sex ratio is male biased (Gowaty and Lennartz 1985).

The helpers are mostly female in the Seychelles warbler. Juvenile females that remain on the parental territory increase parental fecundity when food is plentiful, but reduce fecundity when food is scarce. Mothers bias their sex ratio toward daughters when resources are abundant, but produce an excess of sons when the food supply is low (Komdeur et al. 1997).

It appears that the helping sex is more valuable to a mother because of the return in future fecundity (Trivers and Hare 1976). A few models have been developed (Emlen et al. 1986; Lessells and Avery 1987). I present my own version in this section.

The model is, as always, fully described by the fitness matrix A. Let there be four classes, juveniles and adults for each sex. This model has the same structure as the earlier analysis of juvenile effects on parental survival and fecundity (see *Simplest Models*, p. 147). But here the phenotype of interest is the sex ratio controlled by the mother, and the effects are confined to the influence of daughters on maternal fecundity, given as

$$F(x) = n[1 + 2n(1 - x)bt/\lambda].$$

A mother has $2n$ offspring when she has no help, of which she receives credit for one-half in this diploid model. Her fecundity in the following year is increased by interaction with her $2n(1 - x)$ daughters, where x is the fraction of sons in each brood, and $1 - x$ is the fraction of daughters. Each daughter has an effect b on maternal fecundity. Finally, the probability that a mother was an adult in the previous year and had interactions with her daughters is t/λ.

The fitness matrix is

$$
\mathbf{A} = \begin{bmatrix}
m_0 & m_1 & f_0 & f_1 \\
0 & x^* F(x^*) \alpha & 0 & x F(x) \\
s & t & 0 & 0 \\
0 & (1 - x^*) F(x^*) \alpha & 0 & (1 - x) F(x) \\
0 & 0 & s & t
\end{bmatrix}.
$$

The parameter s is survival from juvenile to adult stage, and t is adult survival. I assume random mating; thus male contributions depend on the normal value of fecundity in the population, $F(x^*)$. A male's expected mating success is the number of females per male, $\alpha = u_1/\hat{u}_1$. The abundance of classes is given by $\mathbf{u} = (\hat{u}_0, \hat{u}_1, u_0, u_1)$, with the hats denoting male classes.

The population growth rate, λ can be obtained from the characteristic equation,

$$
\lambda^2 - t\lambda - 2sF(x^*)(1 - x^*) = 0.
$$

which, as usual, is most conveniently obtained by assigning all offspring to adult females. I assume that population growth is controlled to a constant value λ by adjustment of juvenile survivorship, s.

The class reproductive values are

$$
\hat{v}_0 = s(1 - x^*)/x^*
$$
$$
\hat{v}_1 = \lambda(1 - x^*)/x^*
$$
$$
v_0 = s
$$
$$
v_1 = \lambda.
$$

The solution is obtained in the usual way

$$
\frac{dW}{dg'} = \mathbf{v}\frac{d\mathbf{A}}{dg'}\mathbf{u} = 0,
$$

which yields

$$
u_1 \hat{v}_0 \frac{\partial [xF(x)]}{\partial x}\frac{dx}{dg'_m} + u_1 v_0 \frac{\partial [(1 - x)F(x)]}{\partial x}\frac{dx}{dg'_f} = 0, \qquad (11.7)
$$

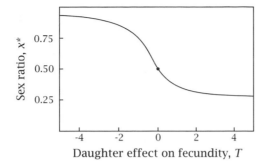

Figure 11.3   Sex ratio produced by mothers when daughters influence future fecundity. The sex ratio, x^*, is the fraction of sons per brood. The effect of daughters is $T = 2nbt/\lambda$, where $2n$ is brood size, b is the benefit per newborn daughter on maternal fecundity in the next year, and t/λ is the probability that a newborn daughter's mother will survive to reproduce in the next year. When b is negative, a daughter reduces maternal fecundity. Note the asymmetry: the male bias created by negative effects is greater than the female bias created by positive effects.

where g'_m and g'_f are the mother's transmitted breeding values through male and female offspring. Relatedness between mother and son is $\hat{r} = dx/dg'_m$, and relatedness between mother and daughter is $r = dx/dg'_f$. The model is diploid; therefore $\hat{r} = r$, and these terms drop out.

Continuing with the differentiation focuses attention on the change in maternal fecundity as a function of sex ratio, x^*, which is

$$F'(x^*) = -nT,$$

where the prime is differentiation with respect to x, and

$$T = 2nbt/\lambda.$$

Substituting into the condition given in Eq. (11.7) yields the solution as

$$1 - 2x^* + T(1 - x^*)(1 - 4x^*) = 0.$$

The solution is illustrated in Fig. 11.3.

The relative reproductive value of male and female offspring in this model is $\hat{v}_0/v_0 = (1 - x^*)/x^*$. This is simply the expected number of mates per male, an effect that arises in all sex ratio models. Thus the demographic context does not alter reproductive values and sex ratios in this particular model. Rather, the marginal gains from sons and daughters depend on how daughters enhance or retard future maternal fecundity. These marginal values are the partial derivatives with respect to x in Eq. (11.7).

Discussions of helpers in birds emphasize demography (e.g., Emlen 1984; Brown 1987). For example, juveniles, living on parental territories, influence juvenile and adult survivorship as well as parental fecundity. One must, in addition, consider the evolution of the helping phenotype in the juvenile. This behavior depends on the opportunities for breeding alone and the coefficients of relatedness among the various actors and recipients. The methods outlined in this book provide the opportunity for formal analysis.

11.7 Multigeneration Colonies

Individuals of some species aggregate into isolated colonies. These colonies last for many generations. Individuals within the colonies compete for resources or cooperate in the acquisition of new resources. Some juveniles remain in the colony; others migrate to join another group or start a new colony. Mating typically occurs within the colony, but migrants can also mate outside their natal group.

Colonial demography poses many interesting problems for sex allocation. The maximum growth rate of the colony depends on the number of females produced in each generation. Colony growth affects the intensity of local competition for resources. In cooperative colonies, group size affects colony efficiency, survivorship and fecundity.

The first models of colonial structure analyzed demography, mating pattern, and competition for resources within colonies (Bulmer and Taylor 1980; Wilson and Colwell 1981; Frank 1986c; van Tienderen and de Jong 1986). The models were then extended to cooperative interactions, in which colony survivorship and fecundity increase as the colony grows larger (Avilés 1986, 1993; Frank 1987a). These cooperative models were based on social spiders, which form multigeneration colonies with highly female-biased sex ratios.

Finally, Nagelkerke and Sabelis (1996) extended the theory and applied their work to the sex ratios of mites. These small arthropods, which have a variety of colonial structures and patterns of sex ratio variation, provide an excellent system for tests of colonial sex ratio theory.

GENERAL FORMULATION

There are many potential demographic and social processes in colonial populations. Several aspects of sex allocation can, however, be expressed by a simple formulation of cycle fitness.

Assume that a colony is formed by a few individuals of generation 0. These generation 0 mothers reproduce to form the first-generation progeny. The sex ratio of an individual mother in generation j, producing the $j + 1$st generation, is x_j, and the average colonial sex ratio in generation j is y_j.

The fitness consequences of individual and colony sex ratios, x_j and y_j, can be expressed by three factors. First, the individual and colony sex ratios in generation j determine the fraction of future progeny that descend from a generation j female. Second, future progeny must be discounted by the population growth rate, λ. Finally, the number of future progeny depends on the fecundity of the colony at each age multiplied by the probability that the colony will survive to that age. These factors combine to give the total reproductive value of future progeny that emigrate to form new generation 0 colonies

$$w_j = f\left(x_j, y_j\right) \sum_{k=j}^{\infty} \lambda^{-j} S\left(\mathbf{y}_k\right) F\left(\mathbf{y}_k\right).$$

The cycle fitness of a female in generation j, with sex ratio x_j and colony sex ratio y_j, is given by w_j. The function f is the fraction of future colony offspring that descend from a female with sex ratio x_j. The survivorship of the colony to produce generation k is $S(\mathbf{y}_k)$, where $\mathbf{y}_k$ is the vector of all colony sex ratios for $j = 0, \ldots, k$. The fecundity of the colony in generation k is $F(\mathbf{y}_k)$.

If the sex ratio produced in each generation is an independent trait, then the direction of selection on the sex ratio in the jth generation is given by dw_j/dg_j'. Because there are three functions, f, S, and F, this derivative has three parts by application of the chain rule

$$\frac{dw_j}{dg_j'} = P_1 + P_2 + P_3$$

where the first part differentiates f, holding S and F constant

$$P_1 = \left[\tilde{r} f_{x_j}\left(y_j^*, y_j^*\right) + \tilde{R} f_{y_j}\left(y_j^*, y_j^*\right)\right] \sum_{k=j}^{\infty} \lambda^{-j} S\left(\mathbf{y}_k^*\right) F\left(\mathbf{y}_k^*\right).$$

The terms f_{x_j} and f_{y_j} are, respectively, the partial derivatives of f with respect to individual and group sex ratio, evaluated at the normal sex ratio, y_j^*. The relatedness coefficient $\tilde{r} = dx_j/dg'_{jx}$ is the slope of mother's phenotype on her transmitted breeding value to offspring. This is equivalent to the relatedness between mother and offspring. I assume a diploid model, with symmetric relatedness to sons and daughters. The coefficient $\tilde{R} = dy_j/dg'_{jy}$ is the slope of a randomly chosen mother's sex ratio, y_j, on the breeding value transmitted to a random offspring in the group, g'_{jy}. This is equivalent to the relatedness between a mother and a randomly chosen offspring in the group.

The second and third terms are

$$P_2 = \tilde{R}f\left(y_j^*, y_j^*\right) \sum_{k=j}^{\infty} \lambda^{-j} S_{y_j}\left(\mathbf{y}_k^*\right) F\left(\mathbf{y}_k^*\right)$$

$$P_3 = \tilde{R}f\left(y_j^*, y_j^*\right) \sum_{k=j}^{\infty} \lambda^{-j} S\left(\mathbf{y}_k^*\right) F_{y_j}\left(\mathbf{y}_k^*\right).$$

The subscript y_j on S and F denotes partial differentiation with respect to y_j. These partial derivatives summarize the viability and fecundity effects on all future generations for a deviation in the sex ratio in generation j.

SOCIAL SPIDER EXAMPLE

The method of the prior section is easy to apply once the biology and associated functional forms are specified. I follow the example given by Frank (1987a, 1273-1274) for social spiders.

I assume that all mating occurs within the colony. Let the fraction of the colony that descends from a particular female, with sex ratio x_j, in a colony with sex ratio y_j, be proportional to

$$f\left(x_j, y_j\right) = \frac{x_j}{y_j} + \frac{1-x_j}{1-y_j},$$

which is a form of the Shaw–Mohler equation introduced in Section 9.1. With this definition, the partial derivative term in P_1 is

$$\tilde{r}f_{x_j} + \tilde{R}f_{y_j} = \left(\tilde{r} - \tilde{R}\right)\frac{1-2y_j^*}{y_j^*\left(1-y_j^*\right)}.$$

The colony is assumed to grow without sending out emigrants up to generation $g - 1$, and then to maintain a stable size and send out emigrants in proportion to $F(\mathbf{y}_k)$ in the following generations. In this particular example, colony fecundity increases linearly with colony size; that is, the number of neighboring females neither increases nor decreases the fecundity per individual. Thus colony fecundity is zero through the first $g - 1$ generations. Colony size and fecundity in the following generations, $k \geq g$, are proportional to

$$F(\mathbf{y}_k^*) = \left[N \prod_{i=0}^{g-1} n (1 - y_i^*) \right] n (1 - y_k^*),$$

where N is the number of founding females in generation 0, and $n(1 - y_i^*)$ is the number of female offspring in the ith generation of a normal colony. The term in square brackets is the size that the colony has achieved during the growth phase, and $1 - y_k^*$ is the number of females produced for dispersal during the reproductive phase. When the sex ratio deviates from normal only in generation j, then

$$F(\mathbf{y}_k) = F(\mathbf{y}_k^*) \frac{1 - y_j}{1 - y_j^*},$$

and the partial of F with respect to the deviant group sex ratio is

$$F_{y_j}(\mathbf{y}_k^*) = -\frac{F(\mathbf{y}_k^*)}{1 - y_j^*}.$$

Colony survival is also divided into two periods. For colony growth, during generation $k < g$, survival in each generation is a function of the current colony size relative to the size of a mature, normal colony

$$\sigma(\mathbf{y}_k) = \delta \left[\frac{\prod_{i=0}^{k-1} n (1 - y_i)}{\prod_{i=0}^{g-1} n (1 - y_i^*)} \right]^{\vartheta},$$

with $\sigma(\mathbf{y}_k) = \delta$ for $k \geq g$. Survival through generation k is therefore

$$S(\mathbf{y}_k) = \prod_{i=0}^{k} \sigma(\mathbf{y}_i).$$

When the sex ratio deviates from normal only in generation $j < g - 1$, then survival in each generation with $k > j$ is

$$\sigma(\mathbf{y}_k) = \sigma(\mathbf{y}_k^*) \left[\frac{1 - y_j}{1 - y_j^*} \right]^{\vartheta},$$

and survival in generations $k \leq j$ is $\sigma(\mathbf{y}_k^*)$. Cumulative survival over generations for $k > g - 1$ is

$$S(\mathbf{y}_k) = S(\mathbf{y}_k^*) \left[\frac{1 - y_j}{1 - y_j^*} \right]^{\vartheta(g-1-j)}.$$

The partial of S with respect to deviant group sex ratio is

$$S_{y_j}(\mathbf{y}_k^*) = -\frac{\vartheta(g-1-j)S(\mathbf{y}_k^*)}{1 - y_j^*}.$$

The partial derivative is zero for generations $j \geq g - 1$.

These terms are used in $dw_j/dg_j' = 0$ to solve for the equilibrium sex ratio produced in generation j. To match the result in Frank (1987a), define $\rho = \tilde{R}/\tilde{r}$ as the value (relatedness) of a mother to a random offspring in the group compared with her relatedness to her own offspring. Using this normalization and solving yields the equilibrium fraction of sons in each generation

$$y_j^* = \frac{1 - \rho}{2(1 + \rho y_j)},$$

where $y_j = \vartheta(g - 1 - j)$ for $j < g - 1$, and zero for $j \geq g - 1$. This result can also be written in ratio form as the allocation to males relative to the allocation to females

$$1 - \rho : 1 + \rho + 2\rho y_j.$$

The terms $1 - \rho : 1 + \rho$ match the standard model of local mate competition and inbreeding (see Section 10.2). The value to a mother for an additional son is discounted by ρ. Here ρ measures the association between the mother's phenotype and the transmitted breeding value of competing mothers who add sons to the local breeding group. The value for an additional daughter is augmented by ρ. In this case ρ measures the association between the mother's phenotype and the transmitted breeding value of neighboring mothers whose sons gain an extra mate.

The additional term is $2\rho y_j$. The term y_j measures the increased survival of the local group provided by an extra daughter in the jth generation. This increased survival is worth 2ρ, the "2" because the increase affects the value of both sexes, and ρ is a measure of association (relatedness) that translates marginal gain into net benefit.

See Frank (1987a) for several other assumptions about sex ratio control and colony fecundity and viability.

12 Conclusions

Now, an observer fresh from Mars might excusably think that the human mind, inspired by experience, would start analysis with the relatively concrete and then, as more subtle relations reveal themselves, proceed to the relatively abstract, that is to say, to start from dynamic relations and then proceed to the working out of the static ones. *But this has not been so in any field of scientific endeavor whatsoever:* always static theory has historically preceded dynamic theory and the reasons for this seem to be as obvious as they are sound—static theory is much simpler to work out; its propositions are easier to prove; and it seems closer to (logical) essentials.

—Joseph A. Schumpeter, *History of Economic Analysis*

My study of the foundations of social evolution was forced on me by my empirical interests. Time and again, when analyzing a problem in the evolution of sex ratio, dispersal, conflict or cooperation, I came across technical difficulties in building a model and understanding the essential process. Eventually it became clear that the difficulties were of the same form. When statics is appropriate, solutions depend on proper measures of value for use in maximization. When statics fail, the dynamics of conflict must be developed.

12.1 Statics

MAXIMIZATION AND MEASURES OF VALUE

The best way to solve static models is to find a measure that is maximized at equilibrium. The proper metric for natural selection is genetic market share (fitness). The problem is to determine how traits influence their own future representation in the population. The effect of changing traits can be measured by three aspects of value.

Marginal value has its typical economic interpretation, the rate of change in components of success as a trait changes.

Reproductive value is similar to the time discount rate for money under the accrual of interest. In biology, discount by time is generalized to any fitness component with a distinct contribution to future market share. Fitness components include different kinds of progeny, nondescendant kin, survival, and fecundity.

Value by relatedness turns out to be two separate phenomena (Chapter 4). The first is the role of correlation among social partners. As the trait of an individual changes, its partners' traits change at a rate given by regression coefficients. Thus we can calculate how a change in individual trait value is associated with changing social environment and how social environment affects an individual's fitness. The regression (kin selection) coefficients summarize statistical information about partners.

The second aspect of relatedness concerns the fidelity of transmission. The value of each fitness component must be weighted by the fidelity by which traits are inherited. This is necessary to measure the way in which current trait value is associated with future market share.

Statics Is a Method

> When a lady visiting [Matisse's] studio said, "But surely, the arm of this woman is much too long," the artist replied politely, "Madame, you are mistaken. This is not a woman, this is a picture."
>
> —E. H. Gombrich, *Art and Illusion*

Perhaps the most common mistake about statics is to assume that it describes an unchanging, stationary world. Statics is, in fact, a method for building models and making comparative predictions (Schumpeter 1954; Samuelson 1983). A model clarifies process and exposes to empirical test hypotheses about relations among variables. The assumption of equilibrium is a gambit—it provides a way forward (Grafen 1991).

12.2 Dynamics

Technical Issues

To solve static models, one constructs a measure perpetually increased by natural selection. The direction of evolutionary change is described by a gradient that is (locally) maximized at equilibrium. Equilibrium provides an opportunity for comparison—how traits change as particular parameters are varied.

The gradient formulation, which provides the direction of evolutionary change away from equilibrium, invites speculation about dynamics. Technical issues intrude, however. A model is a partial analysis, not an accurate description of reality. At equilibrium, the parts excluded often diminish to zero in their effect, so that statics has a relatively secure logical basis near fixed points. Many problems concerning the shift from one equilibrium to another can also be studied in this way (Chapter 5).

Away from equilibrium, one has a description of dynamics that is exact for a partial set of factors, but the excluded set is often significant. One can give up the gambit of perpetual increase for a particular measure, include more factors, and obtain increased realism. The result is often technically difficult, harder to understand, and impractical to test empirically. Hence the enduring role of statics.

CONFLICT AND POWER

Equilibrium analysis fails to capture the essentials of certain problems. Consider, for example, the conflict between two individuals, or two populations. An increase in fitness of one entity requires a decline in fitness of the other. Fitness gradients define the intensity of the conflict—how much one side loses when the other gains. The gradients also specify the equilibria that would occur if one side wholly dominated the interaction.

The crux of the problem is entirely in the dynamics. Toward whose favor will the system evolve? Will power fluctuate, causing continual motion? Or can one side gain permanent control, maintaining its optimum (e.g., Trivers and Hare 1976; Krebs and Dawkins 1984; Frank 1989; Hurst et al. 1996)?

The methods of this book identify the conflict but provide no way to proceed. To study this problem, I turned my attention for several years to the coevolution of hosts and parasites (Frank 1993, 1994c). The conflict is clear. The parasites gain by eating the host; the host gains by preventing attack. The traits are simple—biochemical recognition and physical barriers of defense. Data are relatively abundant because of the biomedical and economic importance of disease.

This is no place to launch a detailed analysis of conflict and its dynamics. But my work did provide one conclusion that is perhaps of general significance (Frank 1997f). I summarize this briefly, both for its potential interest in studies of social evolution and as a counterpoint to the relentlessly static analyses of this book.

The outcome of an interaction between host and parasite often depends on recognition. The host can usually rout a parasite if it can identify invasion. The parasite succeeds if it escapes detection.

The parasite trait may, for example, be the shape of a surface molecule that exists in two forms. One form is invisible to the host, but renders the molecule less useful in its normal, biochemical function. The other form, improved functionally, is more easily detected. Hosts sometimes have matching variability. One molecular form is better at recognizing the visible parasite shape. A different form is better in other biochemical functions or in recognizing other invaders.

Such a system often lacks a stable equilibrium. Fluctuations depend on the frequency of the traits in the host and parasite populations, and the associated rise and fall in the density of surviving hosts and parasites. Here the power struggle is over cost in a matched pair of traits. The less costly the better recognition form is to the host, the more the parasite is favored to carry its costly, invisible form.

Recognition systems rarely depend on a single matched pair of traits when there is conflict. The defender scans many independent channels; the evader flees across this spectrum.

How does the number of channels influence the dynamics of conflict? The struggle occurs over the entire spectrum when there are few channels. There may be fluctuations in frequency of traits and density of defenders and evaders. But the qualitative nature of the battle, defined by the active spectrum, remains constant over time.

The dynamics change as the number of channels increases. Defenders can scan effectively only a subset of channels. Evaders avoid the locally scanned channels; defenders pursue over the spectrum.

The gain and loss of locally novel detection and escape traits control the dynamics. In host-parasite systems with molecular recognition, the biochemistry sets the width of the spectrum and the frequency at which new mutations arise in each channel. Spatial processes influence local loss by extinction and reintroduction by migration. The biochemistry and the demography are the attributes of power that determine the outcome of the struggle. I have supported this view with explicit models of several host-parasite systems and with a wide range of circumstantial evidence (Frank 1994c, 1997f).

Animal communication may often be a coevolutionary arms race analogous to a host-parasite battle (Dawkins and Krebs 1978; Krebs and Dawkins 1984). Conflict in communication occurs when the sender and the receiver of signals have different interests. The type of dynamics suggested by Dawkins and Krebs seems to depend on the width of the communication spectrum. But this idea was not developed explicitly.

Guilford and Dawkins (1991) emphasized that the nature of a signaling arms race depends on the physical properties of the signal and the psychology of the receiver. I would put the matter slightly differently, to match the host-parasite example. The mechanisms of communication determine the costs and benefits of alternative traits within each channel and the width of the communication spectrum. The mechanisms also set the rate for loss and reintroduction of particular traits, and therefore the tendency for evolutionary dynamics to be a game of pursuit across the communication spectrum.

The arms race theory of communication has not been developed by explicit models. It is difficult to see exactly what is required for the theory to work. Analogy to the host-parasite models may provide a broader understanding of the evolutionary dynamics of conflict.

References

Anderson, R. M., and May, R. M. 1991. *Infectious Diseases of Humans: Dynamics and Control.* Oxford University Press, Oxford.

Antolin, M. F. 1993. Genetics of biased sex ratios in subdivided populations: models, assumptions, and evidence. *Oxford Surveys in Evolutionary Biology* 9:239-281.

Aumann, R. J. 1974. Subjectivity and correlation in randomized strategies. *Journal of Mathematical Economics* 1:67-96.

Aumann, R. J. 1987. Correlated equilibrium as an expression of Bayesian rationality. *Econometrica* 55:1-18.

Avilés, L. 1986. Sex-ratio bias and possible group selection in the social spider *Anelosimus eximius. American Naturalist* 128:1-12.

Avilés, L. 1993. Interdemic selection and the sex ratio: a social spider perspective. *American Naturalist* 142:320-345.

Barton, N. H., and Turelli, M. 1987. Adaptive landscapes, genetic distance and the evolution of quantitative characters. *Genetical Research* 49:157-173.

Bennett, J. H. 1983. *Natural Selection, Heredity, and Eugenics, Including Selected Correspondence of R. A. Fisher with Leonard Darwin and Others.* Oxford University Press, Oxford.

Boomsma, J. J. 1989. Sex-investment ratios in ants: has female bias been systematically overestimated? *American Naturalist* 133:517-532.

Boomsma, J. J. 1991. Adaptive colony sex ratios in primitively eusocial bees. *Trends in Ecology and Evolution* 6:92-95.

Boomsma, J. J., and Eickwort, G. C. 1993. Colony structure, provisioning and sex allocation in the sweat bee *Halictus ligatus* (Hymenoptera: Halictidae). *Biological Journal of the Linnean Society* 48:355-377.

Boomsma, J. J., and Grafen, A. 1990. Intraspecific variation in ant sex ratios and the Trivers–Hare hypothesis. *Evolution* 44:1026-1034.

Boomsma, J. J., and Grafen, A. 1991. Colony-level sex ratio selection in the eusocial Hymenoptera. *Journal of Evolutionary Biology* 3:383-407.

Bourke, A. F. G., and Franks, N. R. 1995. *Social Evolution in Ants.* Princeton University Press, Princeton, New Jersey.

Breed, M. D., and Bennett, B. 1987. Kin recognition in highly eusocial insects. In Fletcher, D. J. C., and Michener, C. D., eds., *Kin Recognition in Animals*, pp. 243-285. Wiley, New York.

Brown, J. L. 1978. Avian communal breeding systems. *Annual Review of Ecology and Systematics* 9:123-155.

Brown, J. L. 1987. *Helping and Communal Breeding in Birds.* Princeton University Press, Princeton, New Jersey.

Bull, J. J. 1983. *The Evolution of Sex Determining Mechanisms.* Benjamin/Cummings, Menlo Park, California.

Bull, J. J. 1994. Perspective: virulence. *Evolution* 48:1423–1437.

Bulmer, M. G., and Taylor, P. D. 1980. Sex ratio under the haystack model. *Journal of Theoretical Biology* 86:83–89.

Caswell, H. 1985. The evolutionary demography of clonal reproduction. In Jackson, J. B. C., Buss, L. W., and Cook, R. E., eds., *Population Biology and Evolution of Clonal Organisms*, pp. 187–224. Yale University Press, New Haven, Connecticut.

Charlesworth, B. 1977. Population genetics, demography and the sex ratio. In Christiansen, F. B., and Fenchel, T. M., eds., *Measuring Selection in Natural Populations*, pp. 345–363. Springer-Verlag, New York.

Charlesworth, B. 1978. Some models of the evolution of altruistic behaviour between siblings. *Journal of Theoretical Biology* 72:297–319.

Charlesworth, B. 1994. *Evolution in Age-Structured Populations,* 2d ed. Cambridge University Press, Cambridge.

Charlesworth, B., and Charnov, E. L. 1981. Kin selection in age-structured populations. *Journal of Theoretical Biology* 88:103–119.

Charnov, E. L. 1982. *The Theory of Sex Allocation.* Princeton University Press, Princeton, New Jersey.

Charnov, E. L. 1993. *Life History Invariants: Some Explorations of Symmetry in Evolutionary Ecology.* Oxford University Press, Oxford.

Charnov, E. L., Maynard Smith, J., and Bull, J. J. 1976. Why be an hermaphrodite? *Nature* 263:125–126.

Clancy, D. J., and Hoffmann, A. A. 1996. Cytoplasmic incompatibility in *Drosophila simulans:* evolving complexity. *Trends in Ecology and Evolution* 11:145–146.

Clark, A. B. 1978. Sex ratio and local resource competition in a prosiminian primate. *Science* 201:163–165.

Crespi, B. J. 1990. Measuring the effect of natural selection on phenotypic interaction systems. *American Naturalist* 135:32–47.

Crespi, B. J., and Taylor, P. D. 1990. Dispersal rates under variable patch density. *American Naturalist* 135:48–62.

Crow, J. F., and Kimura, M. 1970. *An Introduction to Population Genetics Theory.* Burgess, Minneapolis, Minnesota.

Crow, J. F., and Nagylaki, T. 1976. The rate of change of a character correlated with fitness. *American Naturalist* 110:207–213.

Crozier, R. H. 1987. Genetic aspects of kin recognition: concepts, models, and synthesis. In Fletcher, D. J. C., and Michener, C. D., eds., *Kin Recognition in Animals*, pp. 55–73. Wiley, New York.

Crozier, R. H., and Pamilo, P. 1996. *Evolution of Social Insect Colonies: Sex Allocation and Kin Selection.* Oxford University Press, Oxford.

Darwin, C. 1859. *On the Origin of Species by Means of Natural Selection.* John Murray, London.

Darwin, C. 1871. *The Descent of Man, and Selection in Relation to Sex.* John Murray, London.

Dawkins, R. 1982. *The Extended Phenotype.* Freeman, San Francisco.

Dawkins, R., and Krebs, J. R. 1978. Animal signals: information or manipulation? In Krebs, J. R., and Davies, N. B., eds., *Behavioural Ecology: An Evolutionary Approach*, pp. 282–309. Blackwell Scientific Publications, Oxford.

Diekmann, O., Christiansen, F., and Law, R., eds. 1996. Special issue: evolutionary dynamics. *Journal of Mathematical Biology* 34:483–688.

Edwards, A. W. F. 1994. The fundamental theorem of natural selection. *Biological Reviews* 69:443–474.

Emlen, S. T. 1978. The evolution of cooperative breeding in birds. In Krebs, J. R., and Davies, N. B., eds., *Behavioural Ecology: An Evolutionary Approach*, pp. 245–281. Blackwell Scientific Publications, Oxford.

Emlen, S. T. 1984. Cooperative breeding in birds and mammals. In Krebs, J. R., and Davies, N. B., eds., *Behavioural Ecology: An Evolutionary Approach*, 2d ed., pp. 305–339. Blackwell Scientific Publications, Oxford.

Emlen, S. T., Emlen, J. M., and Levin, S. A. 1986. Sex-ratio selection in species with helpers-at-the-nest. *American Naturalist* 127:1–8.

Eshel, I. 1996. On the changing concept of evolutionary population stability as a reflection of changing point of view in the quantitative theory of evolution. *Journal of Mathematical Biology* 34:485–510.

Ewens, W. J. 1989. An interpretation and proof of the fundamental theorem of natural selection. *Theoretical Population Biology* 36:167–180.

Falconer, D. S. 1989. *Introduction to Quantitative Genetics,* 3d ed. Wiley, New York.

Fisher, R. A. 1918. The correlation between relatives on the supposition of Mendelian inheritance. *Transactions of the Royal Society of Edinburgh* 52:399–433.

Fisher, R. A. 1930. *The Genetical Theory of Natural Selection.* Clarendon, Oxford.

Fisher, R. A. 1958a. *The Genetical Theory of Natural Selection,* 2d ed. Dover, New York.

Fisher, R. A. 1958b. Polymorphism and natural selection. *Bulletin de l'Institut International de Statistique* 36:284–289.

Frank, S. A. 1985. Hierarchical selection theory and sex ratios. II. On applying the theory, and a test with fig wasps. *Evolution* 39:949–964.

Frank, S. A. 1986a. Dispersal polymorphisms in subdivided populations. *Journal of Theoretical Biology* 122:303–309.

Frank, S. A. 1986b. The genetic value of sons and daughters. *Heredity* 56:351–354.

Frank, S. A. 1986c. Hierarchical selection theory and sex ratios I. General solutions for structured populations. *Theoretical Population Biology* 29:312–342.

Frank, S. A. 1987a. Demography and sex ratio in social spiders. *Evolution* 41:1267–1281.

Frank, S. A. 1987b. Individual and population sex allocation patterns. *Theoretical Population Biology* 31:47–74.

Frank, S. A. 1987c. Variable sex ratio among colonies of ants. *Behavioral Ecology and Sociobiology* 20:195–201.

Frank, S. A. 1989. The evolutionary dynamics of cytoplasmic male sterility. *American Naturalist* 133:345–376.

Frank, S. A. 1990a. Sex allocation theory for birds and mammals. *Annual Review of Ecology and Systematics* 21:13–55.

Frank, S. A. 1990b. When to copy or avoid an opponent's strategy. *Journal of Theoretical Biology* 145:41–46.

Frank, S. A. 1992. A kin selection model for the evolution of virulence. *Proceedings of the Royal Society of London B* 250:195–197.

Frank, S. A. 1993. Evolution of host-parasite diversity. *Evolution* 47:1721–1732.

Frank, S. A. 1994a. Genetics of mutualism: the evolution of altruism between species. *Journal of Theoretical Biology* 170:393–400.

Frank, S. A. 1994b. Kin selection and virulence in the evolution of protocells and parasites. *Proceedings of the Royal Society of London B* 258:153–161.

Frank, S. A. 1994c. Recognition and polymorphism in host-parasite genetics. *Philosophical Transactions of the Royal Society of London B* 346:283–293.

Frank, S. A. 1995a. George Price's contributions to evolutionary genetics. *Journal of Theoretical Biology* 175:373–388.

Frank, S. A. 1995b. Mutual policing and repression of competition in the evolution of cooperative groups. *Nature* 377:520–522.

Frank, S. A. 1995c. The origin of synergistic symbiosis. *Journal of Theoretical Biology* 176:403–410.

Frank, S. A. 1995d. Sex allocation in solitary bees and wasps. *American Naturalist* 146:316–323.

Frank, S. A. 1996a. The design of natural and artificial adaptive systems. In Rose, M. R., and Lauder, G. V., eds., *Adaptation*, pp. 451–505. Academic Press, San Diego, California.

Frank, S. A. 1996b. Models of parasite virulence. *Quarterly Review of Biology* 71:37–78.

Frank, S. A. 1996c. Statistical properties of polymorphism in host-parasite genetics. *Evolutionary Ecology* 10:307–317.

Frank, S. A. 1997a. Cytoplasmic incompatibility and population structure. *Journal of Theoretical Biology* 184:327–330.

Frank, S. A. 1997b. The design of adaptive systems: optimal parameters for variation and selection in learning and development. *Journal of Theoretical Biology* 184:31–39.

Frank, S. A. 1997c. Models of symbiosis. *American Naturalist* 150:S80–S99.

Frank, S. A. 1997d. Multivariate analysis of correlated selection and kin selection, with an ESS maximization method. *Journal of Theoretical Biology* 189:307–316.

Frank, S. A. 1997c. The Price equation, Fisher's fundamental theorem, kin selection, and causal analysis. *Evolution* 51:1712–1729.

Frank, S. A. 1997f. Spatial processes in host-parasite genetics. In Hanski, I., and Gilpin, M., eds., *Metapopulation Biology: Ecology, Genetics, and Evolution*, pp. 325–352. Academic Press, New York.

Frank, S. A., and Crespi, B. J. 1989. Synergism between sib-rearing and sex ratio in Hymenoptera. *Behavioral Ecology and Sociobiology* 24:155–162.

Frank, S. A., and Slatkin, M. 1990a. The distribution of allelic effects under mutation and selection. *Genetical Research* 55:111–117.

Frank, S. A., and Slatkin, M. 1990b. Evolution in a variable environment. *American Naturalist* 136:244–260.

Frank, S. A., and Slatkin, M. 1992. Fisher's fundamental theorem of natural selection. *Trends in Ecology and Evolution* 7:92–95.

Frank, S. A., and Swingland, I. R. 1988. Sex ratio under conditional sex expression. *Journal of Theoretical Biology* 135:415–418.

Gayley, T. W., and Michod, R. E. 1990. Modification of genetic constraints on frequency-dependent selection. *American Naturalist* 136:406–426.

Getz, W. M. 1982. An analysis of learned kin recognition of Hymenoptera. *Journal of Theoretical Biology* 99:585–587.

Gillespie, J. H. 1977. Natural selection for variances in offspring numbers: a new evolutionary principle. *American Naturalist* 111:1010–1014.

Godfray, H. C. J., Partridge, L., and Harvey, P. H. 1991. Clutch size. *Annual Review of Ecology and Systematics* 22:409–429.

Gombrich, E. H. 1969. *Art and Illusion*, 2d ed. Princeton University Press, Princeton, New Jersey.

Goodnight, C. J., Schwartz, J. M., and Stevens, L. 1992. Contextual analysis of models of group selection, soft selection, hard selection, and the evolution of altruism. *American Naturalist* 140:743–761.

Goodnight, K. F. 1992. The effect of stochastic variation on kin selection in a budding-viscous population. *American Naturalist* 140:1028–1040.

Gowaty, P. A., and Lennartz, M. R. 1985. Sex ratios of nestling and fledgling Red-Cockaded Woodpeckers (*Picoides borealis*) favor males. *American Naturalist* 126:347–353.

Grafen, A. 1979. The hawk-dove game played between relatives. *Animal Behaviour* 27:905–907.

Grafen, A. 1985. A geometric view of relatedness. *Oxford Surveys in Evolutionary Biology* 2:28–89.

Grafen, A. 1986. Split sex ratios and the evolutionary origins of eusociality. *Journal of Theoretical Biology* 122:95–121.

Grafen, A. 1990. Do animals really recognize kin? *Animal Behaviour* 39:42–54.

Grafen, A. 1991. Modelling in behavioural ecology. In Krebs, J. R., and Davies, N. B., eds., *Behavioural Ecology: An Evolutionary Approach,* 3d ed., pp. 5–31. Blackwell Scientific Publications, Oxford.

Guilford, T., and Dawkins, M. S. 1991. Receiver psychology and the evolution of animal signals. *Animal Behaviour* 42:1–14.

Haig, D. 1996. Gestational drive and the green-bearded placenta. *Proceedings of the National Academy of Sciences USA* 93:6547–6551.

Hamilton, W. D. 1964a. The genetical evolution of social behaviour. I.. *Journal of Theoretical Biology* 7:1–16.

Hamilton, W. D. 1964b. The genetical evolution of social behaviour. II. *Journal of Theoretical Biology* 7:17–52.

Hamilton, W. D. 1966. The moulding of senescence by natural selection. *Journal of Theoretical Biology* 12:12–45.

Hamilton, W. D. 1967. Extraordinary sex ratios. *Science* 156:477–488.

Hamilton, W. D. 1970. Selfish and spiteful behaviour in an evolutionary model. *Nature* 228:1218–1220.

Hamilton, W. D. 1972. Altruism and related phenomena, mainly in social insects. *Annual Review of Ecology and Systematics* 3:193–232.

Hamilton, W. D. 1975. Innate social aptitudes of man: an approach from evolutionary genetics. In Fox, R., ed., *Biosocial Anthropology*, pp. 133–155. Wiley, New York.

Hamilton, W. D. 1979. Wingless and fighting males in fig wasps and other insects. In Blum, M. S., and Blum, N. A., eds., *Reproductive Competition and Sexual Selection in Insects*, pp. 167–220. Academic Press, New York.

Hamilton, W. D. 1996. *Narrow Roads of Gene Land.* Freeman, San Francisco.

Hamilton, W. D., and May, R. M. 1977. Dispersal in stable habitats. *Nature* 269:578–581.

Hammerstein, P. 1996. Darwinian adaptation, population genetics and the streetcar theory of evolution. *Journal of Mathematical Biology* 34:511–532.

Hardin, G. 1993. *Living within Limits: Ecology, Economics, and Population Taboos.* Oxford University Press, Oxford.

Hasegawa, E., and Yamaguchi, T. 1995. Population structure, local mate competition, and sex-allocation pattern in the ant *Mesor aciculatus.* *Evolution* 49:260–265.

Heisler, I. L., and Damuth, J. 1987. A method for analyzing selection in hierarchically structured populations. *American Naturalist* 130:582–602.

Helms, K. R. 1994. Sexual size dimorphism and sex ratios in bees and wasps. *American Naturalist* 143:418–434.

Herre, E. A. 1985. Sex ratio adjustment in fig wasps. *Science* 228:896–898.

Herre, E. A. 1993. Population structure and the evolution of virulence in nematode parasites of fig wasps. *Science* 259:1442–1446.

Holt, R. D. 1996. Demographic constraints in evolution: towards unifying the evolutionary theories of senescence and niche conservatism. *Evolutionary*

Ecology 10:1-11.

Hurst, L. D., Atlan, A., and Bengtsson, B. O. 1996. Genetic conflicts. *Quarterly Review of Biology* 71:317-364.

Karlin, S., and Matessi, C. 1983. Kin selection and altruism. *Proceedings of the Royal Society of London B* 219:327-353.

Kelly, J. K. 1992. Restricted migration and the evolution of altruism. *Evolution* 46:1492-1495.

Kimura, M. 1958. On the change of population fitness by natural selection. *Heredity* 12:145-167.

Komdeur, J., Daan, S., Tinbergen, J., and Mateman, C. 1997. Extreme adaptive modification in sex ratio of the Seychelles warbler's eggs. *Nature* 385:522-525.

Krebs, J. R., and Dawkins, R. 1984. Animal signals: mind-reading and manipulation. In Krebs, J. R., and Davies, N. B., eds., *Behavioural Ecology: An Evolutionary Approach,* 2d ed., pp. 380-402. Blackwell Scientific Publications, Oxford.

Lande, R., and Arnold, S. J. 1983. The measurement of selection on correlated characters. *Evolution* 37:1212-1226.

Leigh, E. G. 1970. Sex ratio and differential mortality between the sexes. *American Naturalist* 104:205-210.

Leimar, O. 1996. Life-history analysis of the Trivers and Willard sex-ratio problem. *Behavioral Ecology* 7:316-325.

Lenski, R. E., and May, R. M. 1994. The evolution of virulence in parasites and pathogens: reconciliation between two competing hypotheses. *Journal of Theoretical Biology* 169:253-265.

Lessells, C. M., and Avery, M. I. 1987. Sex-ratio selection in species with helpers at the nest: some extensions of the repayment model. *American Naturalist* 129:610-620.

Li, C. C. 1967. Fundamental theorem of natural selection. *Nature* 214:505-506.

Li, C. C. 1975. *Path Analysis.* Boxwood, Pacific Grove, California.

Lindgren, B. W. 1976. *Statistical Theory,* 3d ed. Macmillan, New York.

Lloyd, W. F. 1833. *Two Lectures on the Checks to Population.* Reprint by Augustus M. Kelley, New York, 1968.

Mangel, M. 1994. Barrier transitions driven by fluctuations, with applications to ecology and evolution. *Theoretical Population Biology* 45:16-40.

Mangel, M., and Clark, C. W. 1988. *Dynamic Modeling in Behavioral Ecology.* Princeton University Press, Princeton, New Jersey.

May, R. M., and Anderson, R. M. 1990. Parasite-host coevolution. *Parasitology* 100:S89-S101.

Maynard Smith, J. 1982. *Evolution and the Theory of Games.* Cambridge University Press, Cambridge.

Maynard Smith, J., and Price, G. R. 1973. The logic of animal conflict. *Nature* 246:15-18.

Maynard Smith, J., and Szathmáry, E. 1995. *The Major Transitions in Evolution*. Freeman, San Francisco.

McNamara, J. M. 1991. Optimal life histories: a generalisation of the Perron-Frobenius theorem. *Theoretical Population Biology* 40:230–245.

McNamara, J. M. 1995. Implicit frequency dependence and kin selection in fluctuating environments. *Evolutionary Ecology* 9:185–203.

Michod, R. E. 1982. The theory of kin selection. *Annual Review of Ecology and Systematics* 13:23–55.

Michod, R. E., and Hamilton, W. D. 1980. Coefficients of relatedness in sociobiology. *Nature* 288:694–697.

Motro, U. 1982a. Optimal rates of dispersal I. Haploid populations. *Theoretical Population Biology* 21:394–411.

Motro, U. 1982b. Optimal rates of dispersal II. Diploid populations. *Theoretical Population Biology* 21:412–429.

Motro, U. 1983. Optimal rates of dispersal III. Parent-offspring conflict. *Theoretical Population Biology* 23:159–168.

Nagelkerke, C. J. 1994. Simultaneous optimization of egg distribution and sex allocation in a patch-structured population. *American Naturalist* 144:262–284.

Nagelkerke, C. J., and Sabelis, M. W. 1996. Hierarchical levels of spatial structure and their consequences for the evolution of sex allocation in mites and other arthropods. *American Naturalist* 148:16–39.

Nagelkerke, K. 1993. *Evolution of Sex Allocation Strategies of Pseudo-Arrhenotokous Predatory Mites (Acari: Phytoseiidae)*. Ph.D. Thesis, University of Amsterdam.

Orlove, M. J., and Wood, C. L. 1978. Coefficients of relationship and coefficients of relatedness in kin selection: a covariance form for the rho formulation. *Journal of Theoretical Biology* 73:679–686.

Oster, G. F., and Wilson, E. O. 1978. *Caste and Ecology in the Social Insects*. Princeton University Press, Princeton, New Jersey.

Parker, G. A., and Maynard Smith, J. 1990. Optimality theory in evolutionary biology. *Nature* 348:27–33.

Price, G. R. 1970. Selection and covariance. *Nature* 227:520–521.

Price, G. R. 1972a. Extension of covariance selection mathematics. *Annals of Human Genetics* 35:485–490.

Price, G. R. 1972b. Fisher's 'fundamental theorem' made clear. *Annals of Human Genetics* 36:129–140.

Price, T., Turelli, M., and Slatkin, M. 1993. Peak shifts produced by correlated response to selection. *Evolution* 47:280–290.

Provine, W. B. 1971. *The Origins of Theoretical Population Genetics*. University of Chicago Press, Chicago.

Queller, D. C. 1992a. A general model for kin selection. *Evolution* 46:376–380.

Queller, D. C. 1992b. Quantitative genetics, inclusive fitness, and group selection. *American Naturalist* 139:540–558.

Queller, D. C. 1994. Genetic relatedness in viscous populations. *Evolutionary Ecology* 8:70–73.

Real, L. 1980. Fitness, uncertainty, and the role of diversification in evolution and behavior. *American Naturalist* 115:623–638.

Robertson, A. 1966. A mathematical model of the culling process in dairy cattle. *Animal Production* 8:95–108.

Rogers, A. R. 1993. Why menopause? *Evolutionary Ecology* 7:406–420.

Rose, M. R. 1991. *Evolutionary Biology of Aging.* Oxford University Press, Oxford.

Rosenheim, J. A., Nonacs, P., and Mangel, M. 1996. Sex ratios and multifaceted parental investment. *American Naturalist* 148:501–535.

Rousset, F., and Raymond, M. 1991. Cytoplasmic incompatibility in insects: why sterilize females? *Trends in Ecology and Evolution* 6:54–57.

Samuelson, P. A. 1983. *Foundations of Economic Analysis,* enlarged ed. Harvard University Press, Cambridge, Massachusetts.

Schumpeter, J. A. 1954. *History of Economic Analysis.* Oxford University Press, New York.

Seger, J. 1981. Kinship and covariance. *Journal of Theoretical Biology* 91:191–213.

Seger, J. 1983. Partial bivoltinism may cause alternating sex-ratio biases that favour eusociality. *Nature* 301:1059–1062.

Shaw, R. F., and Mohler, J. D. 1953. The selective advantage of the sex ratio. *American Naturalist* 87:337–342.

Sherman, P. W., Jarvis, J., and Alexander, R. D., eds. 1991. *The Biology of the Naked Mole Rat.* Princeton University Press, Princeton, New Jersey.

Skyrms, B. 1996. *Evolution of the Social Contract.* Cambridge University Press, Cambridge.

Smith, C. C. 1974. The optimal balance between size and number of offspring. *American Naturalist* 108:499–506.

Stephens, D. W., and Krebs, J. R. 1986. *Foraging Theory.* Princeton University Press, Princeton.

Stubblefield, J. W., and Seger, J. 1990. Local mate competition with variable fecundity: dependence of offspring sex ratios on information utilization and mode of male production. *Behavioral Ecology* 1:68–80.

Sundström, L. 1995. Sex allocation and colony maintenance in monogyne and polygyne colonies of *Formica truncorum* (Hymenoptera: Formicidae): the impact of kinship and mating structure. *American Naturalist* 146:182–201.

Sundström, L., and Keller, L. 1996. Conditional manipulation of sex ratios by ant workers: a test of kin selection theory. *Science* 274:993–995.

Suzuki, Y., and Iwasa, Y. 1980. A sex ratio theory of gregarious parasitoids. *Researches on Population Ecology* 22:366–382.

Taylor, P. D. 1988a. An inclusive fitness model for dispersal of offspring. *Journal of Theoretical Biology* 130:363–378.

Taylor, P. D. 1988b. Inclusive fitness models with two sexes. *Theoretical Population Biology* 34:145–168.

Taylor, P. D. 1990. Allele-frequency change in a class-structured population. *American Naturalist* 135:95–106.

Taylor, P. D. 1992a. Altruism in viscous populations—an inclusive fitness approach. *Evolutionary Ecology* 6:352–356.

Taylor, P. D. 1992b. Inclusive fitness in a heterogeneous environment. *Proceedings of the Royal Society of London B* 249:299–302.

Taylor, P. D., and Frank, S. A. 1996. How to make a kin selection model. *Journal of Theoretical Biology* 180:27–37.

Trivers, R. L. 1974. Parent-offspring conflict. *American Zoologist* 4:249–264.

Trivers, R. L., and Hare, H. 1976. Haplodiploidy and the evolution of the social insects. *Science* 191:249–263.

Trivers, R. L., and Willard, D. E. 1973. Natural selection of parental ability to vary the sex ratio of offspring. *Science* 179:90–92.

Tuljapurkar, S. 1990. *Population Dynamics in Variable Environments.* Springer-Verlag, New York.

Uyenoyama, M. K., and Feldman, M. 1982. Population genetic theory of kin selection. II. The multiplicative model. *American Naturalist* 120:614–627.

van Tienderen, P. H., and de Jong, G. 1986. Sex ratio under the haystack model: polymorphism may occur. *Journal of Theoretical Biology* 122:69–81.

Wade, M. J. 1985. Soft selection, hard selection, kin selection, and group selection. *American Naturalist* 125:61–73.

Werren, J. H. 1980. Sex ratio adaptations to local mate competition in a parasitic wasp. *Science* 208:1157–1159.

Werren, J. H. 1983. Sex ratio evolution under local male competition in a parasitic wasp. *Evolution* 37:116–124.

Werren, J. H., and Charnov, E. L. 1978. Facultative sex ratios and population dynamics. *Nature* 272:349–350.

Werren, J. H., and Taylor, P. D. 1984. The effects of population recruitment on sex ratio selection. *American Naturalist* 124:143–148.

Werren, J. H., Zhang, W., and Guo, L. R. 1995. Evolution and phylogeny of *Wolbachia:* reproductive parasites of arthropods. *Proceedings of the Royal Society of London B* 261:55–63.

West, S. A., and Godfray, H. C. J. 1997. Sex ratio strategies after perturbation of the stable age distribution. *Journal of Theoretical Biology* (in press).

Wilson, D. S. 1980. *The Natural Selection of Populations and Communities.* Benjamin/Cummings, Menlo Park, California.

Wilson, D. S., and Colwell, R. K. 1981. The evolution of sex ratio in structured demes.. *Evolution* 35:882–897.

Wilson, D. S., Pollock, G. B., and Dugatkin, L. A. 1992. Can altruism evolve in purely viscous populations? *Evolutionary Ecology* 6:331–341.

Wilson, E. O. 1971. *The Insect Societies.* Harvard University Press, Cambridge, Massachusetts.

Wrensch, D. L., and Ebbert, M. A., eds. 1993. *Evolution and Diversity of Sex Ratio in Insects and Mites.* Chapman and Hall, New York.

Wright, S. 1932. The roles of mutation, inbreeding, cross-breeding and selection in evolution. *Proceedings VI International Congress of Genetics* 1:356–366.

Yamaguchi, Y. 1985. Sex ratios of an aphid subject to local mate competition with variable maternal condition. *Nature* 318:460–462.

Author Index

Subject Index